WHO BANGED THE BIG BANG!

TRANS-SCIENTIFIC MYSTERIES DECODED, SIMPLIFIED!

PRABHA KARAN

ISBN

Paperback 978-969-679-264-2

Objective

Scientists, physicists, mathematicians, theorists, and other ambitious thinkers are accustomed to exploring questions that push the limits of human understanding. This commitment to probing the depths of the universe's and life's mysteries is the source of mankind's greatest accomplishments, such as medical breakthroughs, technological innovations, and philosophical revelations.

And yet, those who spend their days focused on such advanced subjects may find that their thoughts keep returning to even bigger questions—questions that leave the boundaries of human knowledge behind and stretch toward the infinite expanse of the unknown.

When did the universe begin? How did it begin? What is the point of origin of the Big Bang? Why does the universe exist at all? Can the smallest sub-atomic particle be divided further? Why are we born as human beings? What is the purpose of being born as a human? What happens after death? Is there life after death? Questions like these have been asked so often that they've essentially become cliché. But that doesn't change the fact that they are still fundamental to the struggle to understand our universe, creation, ourselves, and the meaning of infinity.

Perhaps they are questions that have confronted you too, in the course of your work and life. This book aims to explore such questions on the macrocosmic scale, discussing the largest phenomena within the observable universe, as well as on the microcosmic scale, down to the Higgs boson and beyond. The purpose of this book is to present known facts and to approach the unknown infinite, the mysteries before the Big Bang, incrementally and methodically.

If you have ever found yourself reflecting on the relationship between your intellectual pursuits and these grand yet essential questions—or if you are curious to *begin* exploring that relationship—please read on. Our goal is to proceed systematically from known physical phenomena toward the elusive concepts that dwell on the fringes of the knowable. We will do so systematically so that gradually—and with your capacity to think, reflect, and contemplate—you will be able to build your understanding of *how the infinite fits into your life and how your life fits into the infinite*

Contents

Chapter 1

Rarest of Humans Probe 'The Infinite'

Though the Earth's human population has now reached eight billion, the opportunity to be born as a human being is the rarest of the rare. Why? Consider the totality of living beings on this planet. The number runs into billions. This includes everything from great mammals to small insects and microbes. If plants are included as well, the number of individual living things becomes breathtakingly large. Compared to this, eight billion is an infinitesimally small number. This is why being born a human is exceedingly rare.

It's even rarer to be born without any physical or mental defects that can prevent or interfere with a fully realized experience of life. And it's rarer still to have the opportunity to get a proper education that introduces life's higher values. Among this small group, how many will ever attempt to probe 'the infinite'? That type of person is the rarest of all beings on planet Earth.

Based on the number of tech universities & scientific institutions across the globe, it's estimated that just about 10 million human beings are either actively or passively pursuing the infinite through objective research and scientific analysis. That's 10 million out of eight billion. And those eight billion are part of an uncountably high number of living beings. To put this in perspective, consider another small group—the current number of millionaires worldwide. In 2024, it was estimated there were about 58 million millionaires globally. So those who pursue and probe the infinite are rarer. Indeed, if you are among this rare 10 million, you are one of the luckiest people in the world.

Within this 10 million, different thinkers have adopted different strategies in their pursuit of the infinite, ranging from the scientific to the spiritual. Some of their research has produced compelling results, but significant gaps remain in our understanding of infinity and creation. When it comes to the infinite, before the Big Bang, there was no consensus amongst the range of thinkers. The purpose of this book is to classify the spectrum of thinkers according to their domain and to explore their apparent pitfalls in the understanding of infinity, the Big Bang, and creation. Our systematic approach will eventually narrow the continuously expanding gap between physicists, scientists, thinkers, researchers, and philosophers.

Before we begin that journey, however, let's take a moment to discuss the spectrum of thinkers in the various fields of study and schools of thought that will come up throughout this book—our cast of characters, as it were. This will help us establish the hard evidence and postulated theories that have brought our collective understanding of the infinite, the Big Bang, and creation to the point it's at today.

Who pursues the infinite?

So, who *are* these rare 10 million individuals? What do they do, and what are their thoughts?

First in line are the physicists, and these include the entire spectrum of astrophysicists and quantum physicists. They are the ones who deal with infinity, one way or another, directly and on a daily basis. They are the students and followers of scientists like Erwin Schrödinger, James Webb, Edwin Hubble, and others who came before or after them. The field of quantum physics is progressively evolving. In 2022, three physicists won a Nobel Prize for their work in entanglement, expanding our knowledge of the quantum relationship between particles and potentially enabling new advances.

Then come the theorists who focus on the more perplexing and extreme phenomena of our universe, the Big Bang, singularity, black holes, and wormholes. Those researching dark matter and dark energy fit in here, too. It is now believed that observable, 'ordinary' matter only represents 5% of all the energy in the universe, with dark matter representing 27% and dark energy 68%. No one knows what dark matter or dark energy actually is, but they do exist, and together they represent 95% of our universe.

Next are the scientists and researchers at SETI or the Search for Extra-Terrestrial Intelligence. This is the organization that probes outer space for signs of intelligent life. The discovery of and communication with extra-terrestrial intelligence would represent a quantum leap forward in our understanding of infinity.

Another important organization is CERN, the European Organization for Nuclear Research. CERN houses the Large Hadron Collider (LHC), a 27-kilometer particle collider used to discover new subatomic particles. The LHC has enabled a number of important discoveries, including that of the Higgs boson. Still, a researcher there has indicated that the size of the LHC is not enough to enable the discovery of *all* subatomic particles. For that, we may need a collider with a diameter of the entire solar system.

Then there are the so-called 'super scientists,' those who seek to go beyond conventional physics with seemingly impossible materials, like super-fluids with zero viscosity or superconductors with zero resistance. The concept of super-consciousness has even been proposed, which would transcend ordinary consciousness. It seems like every discipline has practitioners bold enough to push beyond the accepted boundaries of their field in an attempt to elevate our collective understanding of the rules that govern our existence.

But it's not just the physicists who study the infinite. Many in the scientific community also do so. Medical and biotech researchers, molecular chemists, and mathematicians—these scientists probe the infinite passively. Developers of self-aware humanoids and artificial intelligence (AI) also belong to this group. With trillions of terra-bytes of data, the developers of AI apply themselves to the study of consciousness/awareness and attempt to probe the infinite within our minds. To that end, psychologists who study the conscious, subconscious, and unconscious minds also belong to this rare group.

Beyond the realm of real science, those who theorize about things like time travel and teleportation also have a place in this conversation. Though the science is unreal so far, the desire to transcend time is a form of pursuing the infinite. Similarly, readers and writers of science fiction engage in unreal but completely legitimate explorations of the infinite.

Finally, we also need to include certain non-scientific individuals who fall on the spectrum between the spiritual and the intellectual. These include mystic power seekers, Kundaliners, chakra healers, extreme yoga practitioners, atheists, agnostics, and metaphysicians. This gives an idea of the entire spectrum of individuals who pursue the infinite beyond scientific practice.

Now, let's return to this rare 10 million. As we have seen, they belong to many different fields, practices, and traditions. What do they all have in common? Generally speaking, there are three aspects that they share. First, there is a focus on **space**, either universal, subatomic, or spiritual. Then, there is a focus on **time** and the rules that govern it. Third, there is a focus on **causation** and how it works in systems where cause and effect are cyclical. These three factors unite those who pursue infinity.

Let's now explore some of the answers that the rare 10 million have proposed.

What is infinity?

Many disciplines of science and mathematics are centred on the idea of infinity. But what is it? Is it a phenomenon? Is it numerical? Is it spatial? Is it temporal? Is it a quality? A quantity? Is it even real, or just a useful concept? These are some of the fundamental questions scientists and mathematicians have tried to answer. In some cases, they've come up with compelling explanations. Does this mean they've captured infinity? We don't know enough to say. But let's consider some of the theories they've presented so far.

First, let's see what math has to offer. Mathematics is full of different types of numbers: real numbers, rational and irrational numbers, and some numbers they call transcendental numbers. But beyond all the tangible numbers is the awareness that there is no numerical endpoint. Numbers exist within an infinity of numbers. Mathematicians have even proposed a system for categorizing distinct sets of infinite numbers. These are known as the aleph numbers. So, mathematicians can describe and categorize infinity, but does that mean they were able to capture what that it actually *is*?

Let's next consider Hilbert's paradox of the Grand Hotel. The paradox describes a hotel with infinite rooms that is at full capacity and yet is still able to accommodate an infinite number of new guests. This illustrates how infinite numbers can interact, and yet, once again, is this insight the same as capturing infinity?

Infinity doesn't only exist on unimaginably large numerical scales, it can also be contained within a single number. For example, adding one-half, one-quarter, one-eighth, one-sixteenth, etc. will produce a number that comes infinitely close to one. But it will never reach it. Infinity also exists within recognizable shapes—like a circle, which can be geometrically understood to have infinite sides. Perhaps another name for a circle could be an *Infigon?*

Now, let's consider infinity in calculus. Newtonian calculus played a major role in enabling a revolution in scientific and technological accuracy. However, the limitation of calculus is the limits themselves. Scientists must rely on limits in order to achieve any semblance of scientific perfection. This is what helped Einstein formulate his famous theories on general relativity. He introduced the cosmological constant, and it helped to fix this problem in his work. Some people have proposed that this constant corresponds to the dark energy that constitutes almost 70% of the universe. Understanding dark energy would be a significant advance in the Theory of Everything, which is still a work in progress. When will this theory become the 'Law of Everything'? That would be a major breakthrough in the pursuit of the infinite.

Physics is full of constants. Consider Planck's constant, which defines units of measurement in the context of physical constants. It's a bit ironic that the science that has helped us probe deeper and deeper into infinity is based on made-up finite limits and constants. One day, we may learn what is on the other side of them.

Many phenomena in astrophysics have been explained using infinity. Black holes, for example, contain gravitational singularities that are said to have an infinite density. *"Not even light can escape from a black hole's event horizon"* is a standard cliché you often hear in all documentaries about space! Speaking of the speed of light, astronomical observations of light helped to prove that the universe is expanding and is doing so faster than it did in the past. But what does that expansion tell us about infinity that exists *within* the universe, and that makes up the universe? What does it tell us about the nature of matter, dark matter, and dark energy? We will discuss this at length later in the chapter on creation, where we will talk about *the existence of something that is made of matter and travels faster than space!*

Let's stay a moment longer with the concept of the speed of light though because it helps to illustrate an important point. Often, when we talk about the speed of light, we're talking about stars and other celestial objects. Astronomers remind us that the light we see in the sky has been travelling towards us for perhaps millions or billions of years. This means that we aren't seeing what exists *now*, but rather what existed *then*, at the moment the light originated from its source. This principle is also true of the objects right in front of us. Your visual impression of all the objects around you originated in the past, not the 'present.' You are experiencing successive moments as an ongoing visual phenomenon.

Let's think about this differently, using language and grammar. If you want to describe what is happening in the present, how would you do that? You can use the present tense to describe what is happening *now*. But the moment you say *now*, it has already become *then*; it's already in the past. A more accurate representation would be to use what is known in English as the *present continuous* tense, which encompasses both the past and the future—for example, "I am seeing you." This begs the question, *What is the present? What is now? Can the now be frozen!?*

Of the rare 10 million who attempt to pursue infinity, the scientists—specifically the physicists—have given us most of the major advances in our understanding. However, as we have seen, there is a limit to what they can do. Current scientific and mathematical systems force them to introduce limits and constants into their work. When they run into phenomena that they cannot explain, they often fall back on unknowns. Some may invoke the mysteries of consciousness, the human brain, the possibility of a creator, and the as-yet-undefined dark energy. They accept the explanation that one day, 'someone' will come along and provide the answers.

This book seeks to build on the discoveries and practices of scientists, mathematicians, theorists, and the rest of the rare 10 million. It aims to rectify the shortcomings in their discoveries and present a systematic method to understand infinity, the mysteries of creation, and the Big Bang.

Chapter 2

Scientific Validation Culminates in...?

This book is primarily addressed to those rare 10 million people, the thinkers and theorists engaged in the pursuit of infinity through the rigorous study of the world around us and the observable universe. For you, the belief depends on evidence. Before you can accept any concept or theory, you need to examine the scientific evidence, proven by experimentation, and tested and validated at various limits.

The topic of validation is an interesting one though. Even within the scientific community, different people may use the same or similar evidence to either validate or invalidate a theory, depending on their perspective. Recently, some astrophysicists used evidence obtained from the James Webb Space Telescope (JWST) to suggest that the Big Bang theory had been invalidated. Others, the majority of the community, used different methods and different evidence to suggest that the new telescope had actually *validated* the theory instead. While those who claim that the theory has been invalidated represent a minority, their existence shows the role that individual discretion plays in scientific validation.

Consider Newton's law of universal gravitation as another example. While this law holds true in most circumstances, it fails in extreme scenarios, like those that exist in deep space.

Some people interpret this failure as an invalidation of the law. They might say it has been replaced by Einstein's theory of general relativity. But how accurate is this claim? Not even Einstein claimed that his theory replaced Newton's law. His work was inclusive of Newton's, and Einstein himself openly recognized the limitations of his theories. Incompleteness and/or incompatibility doesn't necessarily mean a theory has been invalidated.

This will become important later. The ideas discussed in this book go beyond Newton's law, Einstein's theories, and even the budding Theory of Everything, but also *include* these concepts. It is not an attempt to invalidate them. There is no choice to make between believing in those previously validated theories and believing in the ideas in this book. That said, these ideas have been validated too. The evidence will be presented for open-minded members of the scientific community to review to their satisfaction.

The process of validation depends not only on the evidence being presented but also on the perspective of the observer. Scientific observation generally depends on two techniques: direct perception and inferences based on data. For example, some theories can be validated simply by what you can see and hear. But human sight only exists within a narrow band of wavelengths and human hearing only within a limited band of frequencies. For phenomena beyond those limitations, we depend on tools like telescopes, X-rays, and infrasonic devices. These instruments capture data, which we can then interpret. In some cases, this interpretation is based on extremely subtle data, like establishing the existence of planets within the 'Goldilocks Zone' of distant stars. In these cases, scientists obtain data, interpret it methodically, make inferences and then validate them.

There is an important lesson to take away here: ***The process of scientific validation depends not only on data and***

evidence but also on a judgment system that exists within the observer. When you are validating a theory, you are actually engaging with something within yourself that is responsible for determining what is and isn't acceptable. In the remaining part of the chapter, we will discuss the nature of your internal system because it is the foundation of the ideas that will be discussed later in this book.

Humanity's defining attribute

Edwin Schrödinger's 'Thought Experiment' involving a cat has become extremely well known. It helped illustrate the principle of *superposition* in quantum mechanics—when particles potentially exist in multiple states at the same time and are only resolved into a single state following external observation. Observation and validation—processes that occur within the individual—are the fundamental components of logical thought and rational belief.

Now, where do these processes occur? What inside of us is responsible for executing these critical functions? They operate within your ***intellect***. Pay special attention to this word because it is an important building block for the ideas that will come later.

So, what is intellect? **The intellect is an aspect of your personality that is responsible for validating your observations. It manifests in the form of a '*single precise thought.*' It is discrete and involves your capacity to think; exercise reason, logic, and comparison; discriminate between what is acceptable and what is not; and analyze and draw inferences from data. It is the single most powerful mechanism of thought within you that enables comparison, decision-making, and scientific validation.** Making the effort to truly understand the nature of the intellect is essential to pursuing and capturing infinity.

To begin with, the dictionary provides useful context for how we can understand the intellect and how we can separate it from similar terms that can confuse us about its true nature.

When talking about someone's mental abilities, most people—even those of you amongst the world's rarest 10 million—will probably refer to the intellect as intelligence or call themselves intelligent. Intelligence is just data collection and storage. **Intellect, on the other hand, refers to the *faculty* of reasoning, thinking, and objective understanding.**

The importance of this distinction becomes clearer if we consider it in the real world. Certain standard measures of an individual's mental power—their intelligence quotient (IQ) or emotional intelligence (EQ)—explicitly use the word *intelligence* in their names. But this can reduce these qualities to a score or a game of points. It obscures the significance of the intellect—the faculty that makes intelligence possible in the first place. What good is a high IQ or EQ if you're ignorant of the actual mechanism behind your intelligence? Focusing on this distinction will help us understand one of the major obstacles to capturing infinity.

Beyond *intelligence,* there are more words that people apply inaccurately when they are actually referring to the intellect. ***Mind*** is one such word. People may say someone has a strong mind, a sharp mind, etc. This is inaccurate and misleading. Remember, the intellect is the human faculty for objective reasoning. The mind, on the other hand, is a flow of thought that houses all your desires and all emotions, positive and negative. It doesn't manifest as a single, discrete thought—and *this* is the distinction between the mind and the intellect.

Love, hatred, anger, jealousy—these things come as a flow of thought. The flow may be controlled or uncontrolled. In either case, it is always a flow rather than a single, discrete thought. **Desires flow as streams of thought toward objects**

or beings that emanate from within one's mind. Again, the distinction between the mind and the intellect will be important in understanding the ideas in this book—and when proceeding along the path toward capturing infinity. *The intellect thinks. The mind cannot think.* Remember this distinction as we move forward.

It doesn't stop there—there are many more distracting misnomers for the intellect. People refer to the human brain as intellect. We know that the brain is a fleshy organ made of gross atoms—the quantum of atoms is the same in someone alive and in a recently deceased body! 'Einstein's brain'—people throw around this epithet to describe someone who has a strong intellect. As an epithet, it may make sense to refer to someone as having Einstein's brain.

People also talk about *genius*. What does it take for a person to become a genius? A genius is a person who has loads of intelligence (data and information) and *also* a strong intellect. That is to say, someone with access to a large amount of data, *and* a powerful intellect that can use that data meaningfully and efficiently. When both traits are combined, this person becomes a genius. But the term *genius* is not interchangeable with *intellect* at all.

Let's return to the topic of EQ since the term is commonly used in popular culture these days. How should we understand what this term even means? The term refers to your ability to understand and respond to emotions—both your own and others. But the term *emotional* obscures the fact that EQ also depends on the intellect. In science, we use data to draw conclusions and validate them; with EQ, emotions are our data, and we use our intellect to interpret them and act accordingly. So EQ is really just another form of the intellect—even though many people attribute it to the mind.

Now, you might be thinking, if intellect is just an alternative word to all the other misconstrued words explained earlier, why are we spending so much time harping on its definition?

We have to because understanding the intellect is essential to everything that comes later in this book. Until we fully acknowledge the distinction between the intellect and terms like intelligence, mind, brains, genius, IQ, and EQ, we won't be able to fully appreciate what the intellect is, and how we are capable of using it.

Now, recognize that it's absurd to use the words intelligence, mind, or anything else in place of the word intellect. The use of **mind** is particularly absurd because, as we will soon see, that word has a very specific real meaning!

All or at least many of you out of the 10 million thinkers use your intellect as your primary source of living. In fact, all of humanity relies on its intellect. But for most people, it is a secondary tool. For all those who think, your success or failure depends on the use of your intellect. Of course, not all of you will use your intellect to the same extent; there is a great gradation among you. This gradation is proportional to the strength of your intellect. *But for the most part, you are using your intellect all the time to make a living—and yet you are unaware of the true significance of this faculty.* This is why it is more important for you than for anyone else to develop a true understanding of it.

Now that we have a working understanding of what the intellect is (and isn't), let's start to look at it along with other parts of our inner selves. To help us understand the relationship between these parts, let's consider the relative hierarchy that exists between plants, animals, and humans.

When we compare plants, animals, and humans, we can observe that plants only possess a physical body—they don't have a mind or an intellect. We next observe that animals have a physical body and a mind, but no intellect. It is only the human being that possesses an intellect in addition to a physical body and a mind.

Now, animal and plant lovers may dispute this, saying that plants have some degree of mind and that animals may possess some intellect. This may be true, but only to an extremely limited extent. For example, there may be some exceptional plants that have very feeble feelings—like the touch-me-not plant or certain carnivorous plants—that exhibit a little bit of movement. But these are clearly exceptions; for the most part, plants have only a physical body. Then, some people may make the case that many animals possess an intellect. Again, to a certain extent, it is true. But it is very rudimentary. It is nothing compared to what even the dumbest human being can achieve.

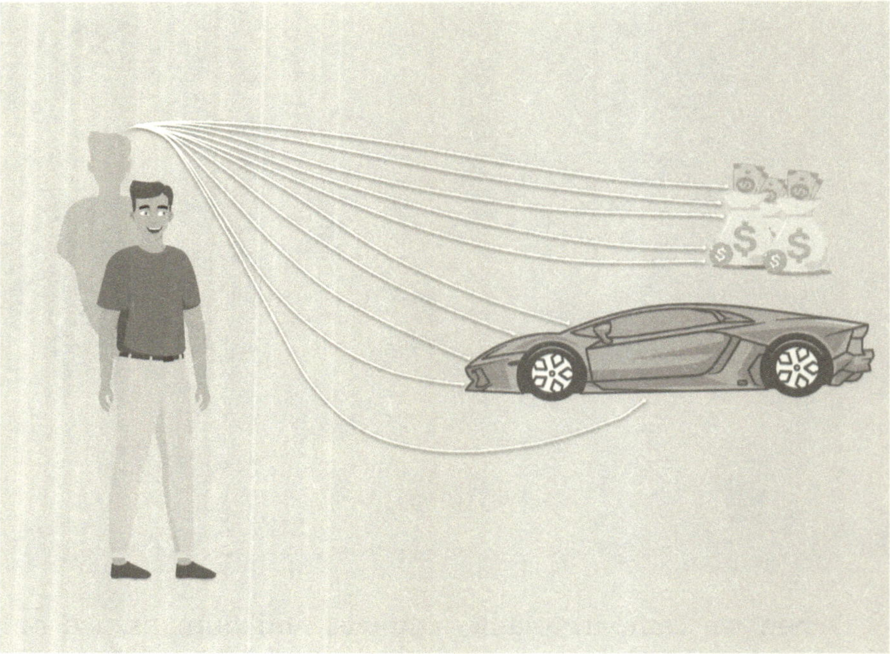

Now, how are we to understand the meaning of **mind** here? Previously, we said that the mind is a flow or pool of thoughts consisting of desires and feelings while the intellect is the faculty to form a single thought. How does this pool flow? In fluid mechanics, there are terms for different kinds of flows, namely laminar (or smooth) flows and turbulent (or tangled) flows. A calm and composed person may have a laminar thought flow—a steady stream of thoughts flowing towards an object. However, as the number of desires increase or if the individual has a cluttered mind, the flow may become turbulent.

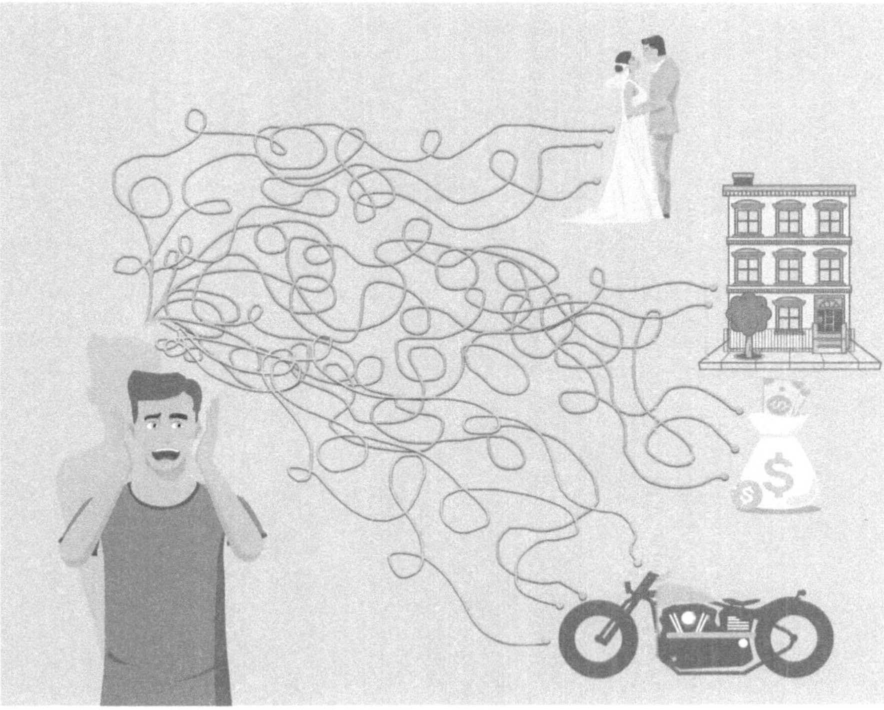

They may have too many thoughts or desires, or their thoughts and/or desires may conflict. This makes the flow of thought more erratic. But, regardless of whether the flow is laminar or turbulent, both streams of thought emanate from the mind.

Apart from the mind and intellect, there are two further components of a person's personality which are depicted here. These are the memory and the ego, which are the deeper parts of one's mind. **Memory houses all the past experiences in the form of thoughts, while the ego declares the identity.** Another way to consider the ego is as a declaration of memories. If a person has zero memories or temporarily loses them, it might be said that the ego is also temporarily lost.

Together, these four components—mind, intellect, memory, and ego—represent the *'subtle body.'* They form our inner selves or inner personalities and exist within us. The intellect gathers intelligence/information/data from memory or external sources. The intellect controls, regulates, and directs the mind, and the mind directs the gross physical body to act. This subtle body is the superior nature of our inner personality, a concept that we will elaborate on in the coming chapters.

Chapter 3

Intellect Ignored and Mind Abandoned

This book is about capturing infinity and exploring & deriving how creation came about, what was before the Big Bang, etc. But to address that, we have to first spend some time exploring the nature and function of the subtle body. This may seem like a detour—but that is exactly the problem! Too many thinkers perform their work without reflecting on or understanding the true nature of the mind and the intellect. This puts a limit on their work and on their chances of actually capturing infinity. *So, we* ***must*** *take the time to explore the subtle body (which is also a subset of creation) to understand our main topic of infinity, creation, and the Big Bang.*

Key functions of the subtle body

Let us first consider the way desires operate within our minds. What effect do desires, and the elimination of them, have on our sense of well-being?

Consider the individual in the first image. The three sets of waves represent his desires, which exist on three levels—physical desires, emotional desires, and intellectual desires. These are represented as waves because the mind's desires do not flow in straight paths, as thoughts in the intellect do. Each wave is one specific desire; together, they create a turbulent flow of many simultaneous desires. This state is common to all human beings. From elite scientists to regular people, everybody entertains many simultaneous, overlapping desires. This image represents the flow of desires in an individual's mind at a given point in time.

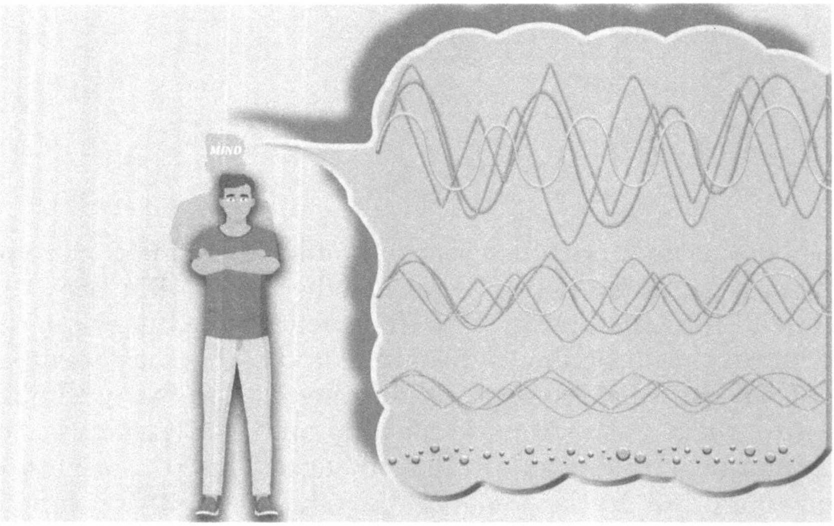

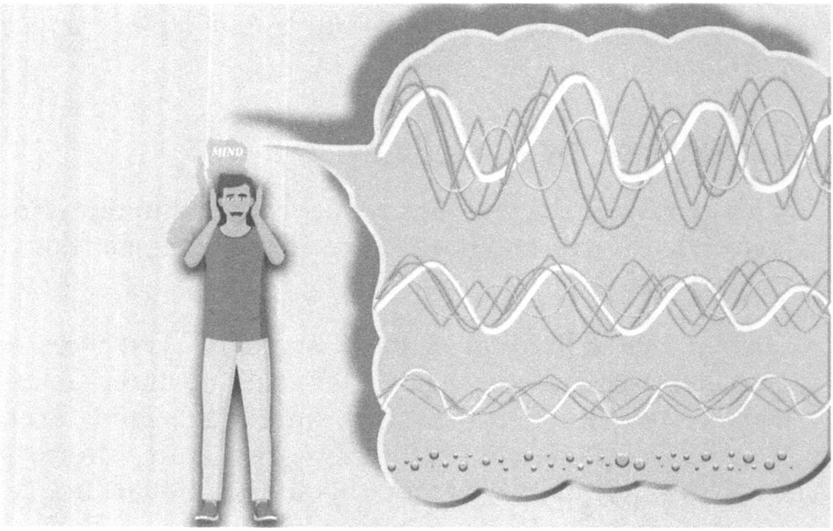

Of course, life is not static, it is dynamic. We experience new desires all the time. In the second image, the same individual has developed three new desires (shown by bright white lines). Now, what does this mean for this individual? In short, these new desires are a source of agitation. Look at the individual's expression in the first image—it's neutral because he's at equilibrium with his existing desires. But his new desires have disrupted that equilibrium because he wants to fulfill them—and has not. These new unfulfilled desires are the source of the newer agitation.

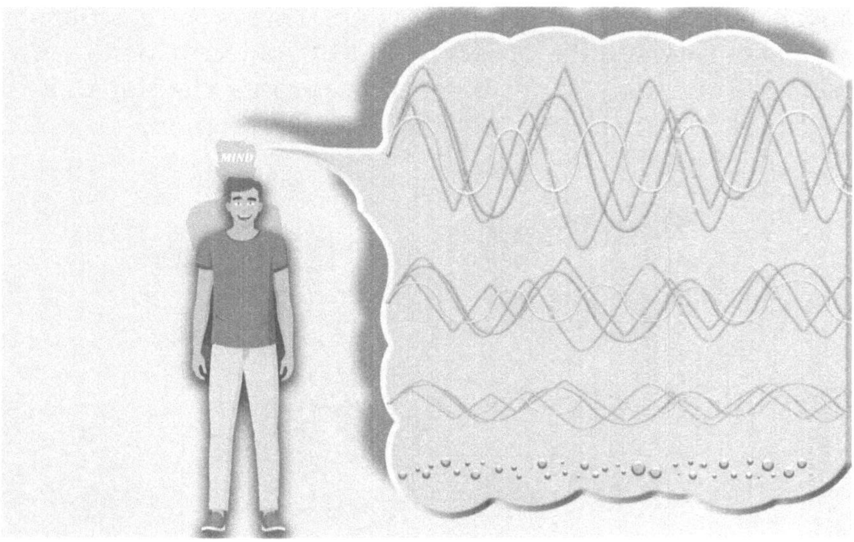

It is human nature to attempt to fulfill our desires, and this individual is no different. He makes use of his physical, emotional, and intellectual capacities to try and fulfill his new desires—and, in this example, he is successful. The white lines have been eliminated, and his flows of thought have returned to their original state. But where does this leave the individual? Compare his expression with the first image. Previously he was neutral, but now he is happy. You've probably experienced this—the sense of relief and happiness that comes from fulfilling desires.

But the important point to note here is that there was no net change in this individual's desires from the first image to the last one. That is to say, it wasn't the number of desires present that mattered, instead, **it was the act of eliminating desire that brought him happiness**. Remember this point, because it will aid your understanding of this topic in future chapters.

The next important topic to discuss as we explore the subtle body is the subject-object relationship. This is the relationship that exists between the observer and the object which is being observed. You may recall from the previous chapter that the intellect is the faculty of observation and validation.

Now, consider the individual highlighted in the image. Everything outside of that person is an object relative to him;

it can be observed and evaluated by his intellect. This extends to other people, to other physical entities, to the Earth, the galaxy, and the entire observable universe. In this instance, the individual is the subject, and everything outside of him represents an object. **When a subject observes or is in contact with an object, it is denoted as an experience**. This is the nature of the subject-object relationship.

Now, it's important to recognize that this relationship doesn't only extend *outward* from the individual to the external object, but can extend *inward* as well. Remember the illustration of the subtle body from the previous chapter—nested within the physical body is the mind, the intellect, the memory, and the ego. So if we think of the mind as a subject, then the gross physical body becomes an object. Further, when the intellect is the subject, the mind can become an object to be observed, evaluated, and experienced. This is an important part of acknowledging the true nature of the intellect; it can observe and relate to objects both in the external world and internally as well. So, whenever we talk about observation and validation, we have to remember to define what we are referring to—what is the subject and what is the object?

Neglecting the intellect

Despite having varied intellectual capacities, many thinkers do not acknowledge or are unaware of the true nature of the intellect as a component of the subtle body. Many people equate the intellect with intelligence, the mind, etc. They believe that they simply *think*, without reflecting on the thing (intellect) that *allows* them to think. This limits their ability to utilize their intellect to its fullest potential. It can also prevent them from ever being able to capture infinity and decipher what is beyond the Big Bang, despite being engaged in the same fields that they actively pursue.

To be a human and ignore the nature of the intellect, at times even refusing to think, is essentially reducing oneself to the level of a microbe! Now, this is not to accuse the reader of being on that level. But by refusing to accept the role of the mind and intellect; by not acknowledging that the intellect plays a major role in your personality; and by not knowing the existence of the subtle body, you are effectively limiting your potential and bringing yourself down to the level of lower forms of life. Even animals possess a subtle body, to an extent. Recall that animals possess a mind in addition to their physical body. If a person remains ignorant of the subtle body, it is equivalent to being at the level of an animal. Widespread ignorance of the subtle body is a major gap in scientific knowledge and training, and it is the main thing holding people back from capturing infinity.

Some may believe that contemplation of the subtle body is irrelevant—that thought is thought, observation is observation, and any discussion of a subtle body is beside the point. The fact of the matter, though, is that the subtle body *does* exist. To deny what is clearly evident is to deny reality. One might as well be the person in the image, standing before an enormous pumpkin and declaring that it doesn't exist. It's an absurdity. The intellect is the tool that scientists, researchers, and thinkers depend on and use daily to earn a living. Denying its nature prevents them from accessing the true capacity of this powerful faculty. This is not a criticism, but rather an illustration of a sorry state of affairs. The top tier of humanity, the world's rarest 10 million thinkers, are generally ignorant of the significance of the intellect.

Open acknowledgement of the role of the intellect could potentially bring about far-reaching advances in the sciences. This could extend up to the cutting-edge fields of research into neutrinos, dark matter, dark energy, infinity, and creation. These specific concepts will be discussed in more detail in the chapters on creation and consciousness, which describe how the ideas contained in this book can be applied to these complex, novel fields. An incomplete application of the intellect is the only thing standing in the way of breakthroughs in experimental validation of several other advanced subjects.

Now, for the reader who is eager to leave these limitations behind, this book will enable you to do exactly that. Throughout your life, you have spent so much time and so many resources on gaining skills and positioning yourself to be the best in your field. By concentrating on the lessons contained in this book, you will finally put yourself in a much better position to comprehend and even capture infinity and decipher concepts beyond the Big Bang. But it's important to understand what *concentration* means here. Many will promptly define concentration as focus. What does focus mean? Concentration!

Concentration has a specific meaning that relates to the subtle body. Understanding this will help you to relate to it more deeply and commit to it more meaningfully.

Successful concentration depends on an interaction between the intellect, the mind, and the gross physical body. To truly concentrate on something, you must first have an intellectual desire for that thing in your mind. When we desire to fulfill a higher intellectual principle, we can give that desire an elevated name—we call it a *goal*. Now that we have established our goal, we can use our intellect to achieve it. ***Concentration is the capacity of the intellect to guide the mind to continue its present activity, without allowing the mind to slip into past experiences or future expectations to accomplish a clear goal.*** It is a collaboration between the mind and the intellect, geared around remaining committed to an ongoing activity. This is the form of concentration that you must bring even to this book in order to derive the value of the lessons contained in it.

Currently, most of the rare 10 million thinkers are not committed to using their intellect in the way described here. In their pursuit of infinity, many of the thinkers may get stagnated, and their intellect gets stuck as it were at a point in their pursuit. They do not have access to a direction, methodology or guidance to further their cause to capture infinity. Going forward, this book will present systematic lessons to guide them toward pursuing and capturing infinity and understanding the Big Bang.

What to expect from this book

So, what will these lessons consist of? The next chapter will build on the subject of goals and discuss them in detail. It will focus on the goal of capturing infinity, the ultimate goal a human being can entertain. It will comprise an in-depth analysis of the nature of human goals.

Next, we will discuss the logic behind capturing infinity. The chapter will contain a concise description of the reason why this should be a goal at all. Later we will go into a thorough explanation of *how* this can be achieved, but it is important to have a clear sense of the rationale first; it will help to organize our ideas going forward.

Next, we will discuss human composition. This will position the subtle body, which has already been introduced, in the context of the *gross physical body.* Another concept, the *causal body*, will be introduced. Altogether, these chapters present an in-depth analysis of the relationship between the components of human composition and the role each is responsible for playing.

Following this discussion of human composition, we will discuss the five layers of the human personality. We will then also explore the three states of human experiences—the waking state, the dream state, and the deep sleep state.

At this point, we will be prepared to discuss infinity itself. We will study infinity and consciousness in detail together. Note that these terms will be explicitly defined at this stage. As we saw, terms like *intellect* and *mind, infinity* and *consciousness* are thrown around carelessly, too. But these chapters will present a nuanced definition of these all-important terms.

Then, we will delve into the next important topic—creation. We will address the Big Bang, dark matter, dark energy, and creation itself. The primary focus of this section will be the question of what came before creation i.e. before the Big Bang. This question is of critical importance in the context of capturing infinity.

At this point, we will *almost* be ready to talk about how to capture infinity. But first, we will spend some time discussing how to become eligible to pursue and capture it. While the rare 10 million do represent the most eligible among us, this does not dismiss everyone else from eligibility. Indeed, *all* humans are eligible for capturing infinity. But there is a distinct threshold for being able to do so. The yardsticks go beyond intellectual capacity, extending to the fitness of the mind and body as well.

These yardsticks will be discussed in detail in those chapters—but in short, one must have the proper intellectual capacity, the purity of mind, and the physical fitness to be able to pursue infinity. They must be able to make a serious commitment to pursuing this goal in order to achieve it.

Assuming one is eligible, one must then prepare oneself for the attempt. The next chapters will discuss the process for doing so. It includes certain preparatory steps, and a few important mandatory intermediary practices, culminating in comprehending infinity.

Having understood the importance of the preparatory practices, we will move on to capturing infinity. What *is* capturing infinity? What happens to you when this is achieved? What happens *after* it has been achieved? All this will be addressed in detail. Then, finally, we will conduct a summary that synthesizes everything presented throughout this book. It is critical to maintain your concentration throughout so that you can derive the most meaning from this experience and position yourself to capture infinity.

Chapter 4

Human Goals

In the previous chapter, we defined *goals* as the products of desires that have been given a clear direction. They have been elevated above the turbulent thought flows of desire in general and this allows them to be pursued and fulfilled by applying intellect. This process of defining goals in the mind and deploying the intellect to achieve them is the nature of concentration. In this chapter, we will define the several types of human goals, discuss the relationship between them, and describe their limitations.

In our needs, we are like plants and animals

In the second chapter, we discussed the fundamental differences between plants, animals, and humans. Recall that plants only have a physical body, while animals have a body and a mind; and only humans have a body, mind, and intellect. All the same, there are certain common, instinctive features shared by all three classes of living beings. These are their *needs*.

Chief among these shared features is the need to eat, for sustenance. From carnivorous plants that entrap and consume insects to photosynthetic ones that survive on light and carbon from the atmosphere, all plants consume some form of fuel. Likewise, from tiny microbial animals to huge carnivores, animals spend most of their waking time roaming around looking for food. And human beings, of course, must eat as well or else they will perish. So, all plants, animals, and human beings must eat. This is the first shared feature among these three classes of living beings.

The second common feature between them is the need for rest. Plants have cycles of rest that occur daily and seasonally. Even more than eating, animals spend the majority of their time sleeping. Human beings also sleep. In fact, both in early infancy and in old age, humans spend nearly all their time sleeping—it's that powerful a need. So the need to rest or sleep is another common feature between these three forms of life.

The third is the need to reproduce or create progeny. Plants, animals, and human beings are constantly producing babies. All forms of life share a biological urge to ensure that their species flourishes. Plants produce innumerable seeds and have evolved ingenious ways of spreading them far and wide. Animals mate and reproduce year after year and will viciously defend their young or even sacrifice themselves to ensure their survival. And humans are constantly reproducing as well, filling the planet with eight billion people.

The fourth common feature seen in all plant, animal, and human life is a sense of insecurity, and the need to seek safety. Plants have evolved defense mechanisms to defend themselves from threats; animals too exhibit fear and have developed fight-or-flight responses to protect themselves. Similarly, humans also exhibit fear. This is an instinctive response to the world around us. The degree to which this manifests in humans has to do with our relative selfishness—the more selfish you are, the greater the sense of insecurity and fear.

These four features—the need to eat, rest, reproduce, and seek safety—are common in all forms of plant, animal, and human life.

Goals set us apart from other forms of life

Beyond these shared features, there is one attribute that distinguishes us from plants and animals—the intellect. This is the extraordinary, defining characteristic of human beings. It allows us to devise goals and systematically plan for our futures. Plants and animals can't set goals; only humans can. This is what sets us apart from other forms of life.

Throughout life, human beings experience countless desires—physical, emotional, and intellectual. This means we have countless opportunities to devise goals and make plans. We also have the opportunity to use our intellect to execute those plans and achieve those goals. **Despite the vast array of potential goals, all human goals can be broadly classified into one of the following four categories.**

Goals of security

The first category includes goals that have to do with security or survival. There are certain resources that a human being needs in order to survive and feel secure, which can remove the sense of insecurity and fear. Goals that have to do with meeting these basic needs fall into this category. What are these basic needs that must be met? They are food, clothing, shelter, and health.

Now, the quantity and/or quality of the food, clothing, shelter, and health that a person requires can vary substantially from one individual to the next depending on their station/status in life. Nonetheless, goals of survival involve satisfying these four basic needs. For example, a poor person may survive with one meal a day, while others may require three meals a day. But in both cases, food is essential for survival. Similarly with clothing—a poor person needs only what will protect the body from adverse temperatures and conditions, while others may need more. But in all cases, clothing is a necessity.

A shelter is the third necessity, whether it is simply a place to get out of the elements or a more elaborate home to raise one's family. No matter whether it's a small tent or a villa, the goal of securing shelter is common among all people and represents the third of our basic needs. The fourth of these goals is health. The most vulnerable people may simply seek to avoid serious illness and bodily harm, while others may aim to prevent illness and try to live healthy lifestyles. Wealthy people, on the other hand, may set goals that include treatments and therapies that are out of reach for almost everyone else. But, as with the three, health also remains a necessity for all people. All over the world, needs that involve these basic necessities form this first type of human goal—the goal of security.

Note that these goals may not be limited to oneself. It can also extend to people that one cares for. While the poor may struggle to meet their present needs of food, clothing, shelter, and health—the middle class and wealthy may be able to secure these needs for themselves and then turn their focus to securing them for their children or relatives. In the same way, after meeting present needs, one may look to secure future needs as well, for example, through health insurance or retirement plans. These types of goals—securing the needs of others and securing one's own needs in the future—are also types of goals of security.

Goals of comfort

Once someone has fully met their goals of security, they can turn to another type of goal—comfort, luxury, and entertainment. These are the things that human beings crave beyond mere survival. As with goals of necessity, goals of comfort have gradations that vary from one person to the next. But these goals always have to do with improving one's standard of living. People want to travel from economy to business class. From riding a scooter, people want to drive a car, and from a normal car, people aim for a luxury car. From a one-bedroom apartment, people want to move to a three-bedroom house, and from there to a mansion. Any goal that goes beyond basic survival falls into the category of comfort, luxury, and entertainment.

Since these goals depend on satisfying one's goals of security first, they are not common to all people. Instead, they are available only to those who enjoy more privileged positions in society.

Goals of invisible wealth

The third category includes goals that have to do with earning *invisible wealth*. What does this mean? It has to do with receiving rewards or good fortune for seemingly no reason. But how can this exist, in a world in which nothing comes for free? And beyond that, how can this be a goal?

People from all walks of life will tell you that nothing comes for free, and yet some win the lottery, strike a fortune in business, or have some other unexpected windfall. These rewards come to people who possess invisible wealth. And the truth is, there *are* certain actions that you can take in this life that will help you earn it. How else we can account for the experiences of children born into rich, powerful, and influential families? What did those children do to be born into such an environment? Children don't have any say in which family they are born into—at least, they don't in *this* life. But if that child had accumulated invisible wealth in a past life, that could influence their birth in this one. Now, you may be wondering why a book that promised to take a systematic, scientific approach is now introducing the premise of life after death. This is a big question—too big to go into detail here. For now, simply know that the theory of reincarnation will be discussed in further detail in the chapter on causality later in this book.

However, because the readers of this book are scientists and thinkers who wouldn't accept anything illogical, even temporarily, we'll briefly touch on the subject. The existence of life after death rests on the law of conservation of matter and energy. Energy and matter can neither be created nor destroyed but can only be transformed from one form to another. The relationship between these laws and human desires is the key to understanding the phenomenon of life after death.

Desires are a form of energy that is manifested as flows of thought. Again, thoughts will be discussed in depth during the chapter on causality. For now, suffice it to say that thoughts and desires are streams of energy. Now, we have discussed how a person houses countless desires over the course of their existence. When a person dies, what happens to all those mental energies? We understand that, upon death, a body is buried or incinerated, returning the gross matter to gross matter. This is in accordance with the law of conservation of matter. Similarly, the law of conservation of energy prevents those accumulated mental energies, expressed as thoughts and desires, from disappearing into nothing. The intellect must accept that those thoughts are conserved in one way or another. This is one indication that there could be life after death.

Now we have to ask ourselves whether we have ever seen any evidence to support the existence of this phenomenon. One potential proof may be found in the behaviors babies sometimes exhibit while sleeping. Even the youngest infants can be observed to be smiling while they sleep. The baby is sleeping, so we know there is no immediate cause for this smile, other than a dream. Dreams are nothing but past experiences stirring in the mind, so it can be inferred that this baby's expression of happiness is a manifestation of some past incident. However, babies are not mature enough to record experiences of their current life, so this smile must come from a dream based on experiences from a past one. This is the first indication that there could be life after death, and before birth.

A second indication can be observed when the baby is three or four years old. At this age, children begin to exhibit specific inclinations, talents, and preferences. Where did these come from, all of a sudden? They already exhibit a certain nature before they develop their intellect. This is particularly evident with child prodigies—musicians, chess players, math geniuses, etc. Where did these talents originate from? In the same way that thought energy cannot disappear into thin air upon death, neither does it spring into existence from nowhere upon birth. This is another indication that there could be life after death. But, again, this subject will be discussed in more detail in the chapter on causality.

Now, let us return to the goal of earning invisible wealth. The cumulative sum of actions taken in the past lives, and in the earlier periods of the current life, generates invisible wealth, which can manifest as rewards in the present. Therefore, performing actions that may earn invisible wealth, either in this life or the next, is the third goal.

Defects in these three goals

These are the first three human goals—goals of security, comfort, and invisible wealth. While these goals may be focused on specific outcomes, ***the underlying motive behind them is the same. Human beings always want to be a little happier than they are right now. All the three categories of goals contribute to increasing our happiness. However, happiness doesn't always immediately result from pursuing or even achieving these goals. This is because they carry inherent defects.*** Let us explore the nature of these defects.

First, the pursuit of goals in these three categories is always mixed with some measure of ***pain and sorrow***. We struggle to achieve our goals, and we take pains to preserve what we have achieved. Thus, even when we accomplish what we set out to do, we still experience pain and sorrow. **Second**, there is always an element of ***dissatisfaction*** inherent in these three categories of goals. What happens when a poor man becomes a little richer? Is he satisfied? No, he wants to gain more *riches*. As we discussed earlier, there can be gradations in what gives one a sense of security or comfort. This is because dissatisfaction is inherent in these three types of goals. Dissatisfaction is always present; even the richest person in the world still wants to be a little richer.

A **third** inherent defect is the sense of ***bondage*** and attachment that forms between us and the objects that help us satisfy our goals. This is rooted in selfishness—in the selfish love we feel toward the objects that we acquire. As with our standards for security and comfort, there are gradations in selfishness among people. But selfishness itself is inevitable in most human beings. Greater selfishness creates greater attachment, and thus more misery; less selfishness, then, leads to less misery. But, in both cases, some degree of misery and pain results from the attachment that develops between us and the things we accomplish.

Fourth Goal—Capturing INFINITY

Now, beyond these three goals ***is a fourth one—the goal of capturing INFINITY.*** This is the ultimate goal any human being can aspire to achieve, and all humans are vicariously pursuing this fourth goal all their lives. Before covering this goal, though, let us quickly recall one key topic that will provide important context for the remainder of this discussion.

What is the nature of a single desire? ***A desire is a steady, continuous stream of thought emanating from the mind towards an object or a being.***

Now, how are goals formed? They are formed by the intellect to help us fulfill our desires. The intellect is necessary to turn and consolidate our desires into goals.

To capture infinity one requires a strong intellect and access to more intelligence than the other goals. One must possess a general knowledge of causation and a desire to probe further into the origins of the world and other mysteries of life in order to truly want to pursue and capture infinity.

With that said, let us move on to a discussion of this fourth goal—capturing infinity.

Goal of capturing infinity

Very few people engage in the pursuit of infinity because it takes a strong intellect and a great deal of intelligence to even set this goal, let alone achieve it. As rare as pursuing the goal of comfort is, this fourth human goal is even rarer; it is only pursued by the rare 10 million.

In the next chapter, we will discuss the logic behind capturing infinity, and further on we will discuss how to actually go about it. In the meantime, let us focus on the benefits that come from applying oneself seriously to this rare fourth goal. This will help us understand its importance relative to the other three goals.

The first benefit you will get from the pursuit of this goal is freedom from ignorance of what infinity is. Even if you are knowledgeable about all aspects of the observable universe, there are still many pockets of ignorance that nobody has been able to illuminate yet. These pockets include things like what caused the Big Bang, creation, causation, dark matter, dark energy, and other newer discoveries that have not been explained yet. These will be systematically removed, step by step, as you pursue the goal of capturing infinity.

The second benefit of pursuing this goal is freedom from all doubts. This flows from the first benefit because ignorance is the breeding ground of questioning and doubt. As you pursue infinity and begin to lose your ignorance, those questions and doubts will also be cleared away. This will not happen instantaneously but will occur during the pursuit of your study.

Doubts tend to lead to misunderstandings and misconceptions. This is because we tend to fill in the gaps in our understanding with patterns of thought based on the information we have already gathered. Such misconceptions are caused by doubts, and doubts are caused by ignorance. With methodical study while pursuing infinity, these misconceptions will be systematically removed.

A third benefit is the reduction of the suffering involved in the pursuit of your goals. This is a somewhat indirect benefit, but it becomes more direct the longer you pursue the goal of capturing infinity. Because it is a noble goal, one must be relatively unselfish to even attempt it. Thus, attempting this goal is an indication that you are a comparatively unselfish person. This means that, right from the start, you will be less affected by the suffering that comes from the inherent defects of the first three human goals. In this respect, this benefit is initially an indirect result of pursuing infinity. Over time, however, the benefit becomes more direct. This is because, as you continue to pursue an unselfish goal, your knowledge becomes wider, and you become wiser. As you become wiser, the effects of those inherent defects will continue to subside. You will still be affected by them, but compared to others you will be much better off.

For one thing, you will be affected by these defects much less frequently. You will also experience them less intensely, and you will recover from them much more quickly. Further, as you become less selfish, you will experience less fear; this means that you will become more courageous. What does it mean to lose fear and gain courage? When you are fearful, your mind projects multiple imaginary adverse effects and expends effort examining each of them. This leads to unnecessary mental suffering. Causes do have tangible effects, but a fearful, selfish mind projects additional imaginary effects and suffers for each of them. Increasing courage, along with increasing knowledge, reduces this form of suffering. Even if you still have some fears, you will be able to face them because the strength of your intellect will be able to confront and understand them.

A fourth benefit is the increased efficiency of your intellect. The pursuit of infinity will increase the strength of your intellect. You will be able to concentrate better, be able to direct your intellect to engage only in the present activity, and not allow your mind to slip back into the past or run ahead into the future. In doing so, you will make more effective use of your time, and you will be happier in your work.

Before we move on to talk about the logic behind capturing infinity in the next chapter, it's worth pointing out that pursuing this goal doesn't mean that you don't pursue the other three goals. Those simply become secondary goals. You can still engage in them, but your success or failure at achieving them will not have an adverse impact on your everyday life. Because of the benefits involved in pursuing this goal, you will be relatively free from the pain, bondage, suffering, and general dissatisfaction involved in pursuing the other goals. Note that you don't have to wait until after you have captured infinity to enjoy these benefits, rather, you gain them during the pursuit itself. Thus, your commitment to the goal of capturing infinity not only serves your pursuit of that goal but the other goals as well.

With that said, let us move on to discuss the logic behind capturing infinity. Though we're not yet ready to actually discuss the nature of infinity or the methods for pursuing and capturing it, it will be beneficial to at least cover the logic behind it. As someone who thinks, you are already prepared to understand this, even though there is still so much more to explore on this subject.

Chapter 5

Logic Behind Capturing Infinity

In order to understand the logic behind capturing infinity, it will help to first look at it in the context of the three other categories of human goals. Recall what those goals were: the goal of security (food, clothing, shelter, and health); the goal of comfort, luxury, and entertainment; and the goal of invisible wealth. These goals are the products of our desires. And while human desires are innumerable, all human goals fall into one of these three categories along with the fourth goal—capturing infinity. To refer to this as the 'fourth goal' is something of an inaccuracy, because in reality it is the noblest goal of all and should be the primary goal. However, only the rarest of people even attempt to pursue it, let alone achieve it.

While the three former categories all contain inherent defects, the fourth has none of these. On the contrary, it includes a number of inherent benefits. These are freedom from ignorance regarding the nature of infinity, freedom from all doubts, a reduction of the suffering that comes from the inherent defects of the other goals, and an increase in the efficiency of your intellect and your actions. And then, finally, you will capture infinity. This represents the crowning benefit and is the ultimate achievement of this noble goal.

While later chapters will discuss the actual strategy for achieving this goal, in this chapter we will focus on the logic behind it. This will help to establish a common understanding of the true purpose of this effort.

But first, a brief aside. In the previous chapter, we touched on reincarnation as it relates to invisible wealth. This may have some readers wondering if this involves the concept of karma. In short, yes—this book does invoke the concepts of karma and reincarnation, in addition to the practices of yoga and meditation. However, these words have been highly commercialized, particularly in the West, which has diminished some of their meaning. The reader should be careful to not let preconceived notions about these terms creep into their understanding of them. Instead, simply follow along as they are presented in this book.

In fact, the reader should try to avoid getting carried away with parallel thinking on any subject, be it yoga, karma, or any of the scientific subjects discussed here. Because the readers will likely have above-average intellect, they will probably have ideas and opinions about many of the topics discussed here. It may be tempting to start pulling in existing knowledge on a given subject whenever it comes up. This is a natural tendency but it should be avoided during these lessons because it will interrupt the book's systematic presentation. In order to get the most out of this experience, try to check the impulse to draw on your previous knowledge.

With that disclaimer out of the way, we are ready to proceed to our discussion of the logic behind capturing infinity.

The happiness formula

We have established that the mind is the seat of human desires and that one desire equals the flow of thought toward one object. That object could be an idea, a physical item, or a living being, such as another person. We have also established that desires can flow in a steady, laminar flow, or a tangled, turbulent flow. Because we entertain many desires simultaneously but only have so much time to fulfill them, our thoughts generally go helter-skelter.

Flows of thought are the building blocks for the entire subtle body. So far we've talked about how they function in the mind and the intellect, but they also operate in the memory and the ego. This will be discussed at length in subsequent chapters. Together, the four components of the subtle body use flows of thought to drive the physical (or gross) body to act. Again, the composition of the gross body will be discussed in future chapters.

Now, as we discussed in the previous chapter, why do humans want to achieve goals? It's an attempt to achieve happiness. Regardless of the goal or goal category, the underlying want inherent in all goals is the desire to become happier—even just a *little* happier than right now. But how can we ever know how happy we actually are? Is there any way to quantify this? Happiness and desire may seem too abstract to quantify. However, we have established that each goal represents the attempt to fulfill one desire. Knowing this, is it possible to devise a formula for measuring one's happiness?

As we saw in the third chapter, we become slightly happier whenever we fulfill a desire (even taking into account the inherent defects of our goals). Thus, we can propose the following: *at a given reference point in our life,* **our happiness can be measured by dividing the number of our desires that have been fulfilled by the total number of desires we have entertained.** Expressed as a formula, we get the following:

$$\textit{\% of happiness} = \frac{\textit{Total number of desires fulfilled}}{\textit{Total number of desires entertained}} \times 100$$

Let's explore how this formula works.

First, how do the numerator and denominator in this formula behave? For any human being, the denominator here is dynamic. It continuously rises and falls as we entertain new desires and forget about old, unfulfilled ones. The numerator, on the other hand, only increases. As we fulfill our goals, this number continues to grow. Most people attempt to increase their happiness by increasing the numerator i.e. by achieving more and more goals and satisfying more and more desires. But does this really work?

The problem with this approach is that, even as one fulfills their desires, new desires are born. After all, humans always want to be happier than they are now. Even as you increase the numerator, you may also be increasing the denominator, which may result in your happiness stagnating or even going down, despite having satisfied some goals.

Is there a better way? Mathematicians reading this may have already considered an alternative—attacking the denominator instead. Because the numerator remains fixed, by reducing the number of desires entertained, one can effectively increase their overall happiness without fulfilling any new desires. Later in this book, we will discuss specific methods for reducing desires that do not involve pain, sacrifice, expending resources, or anything else that may cause suffering. Of course, new desires will constantly crop up, because the flows of thought that correspond to desires cannot always be managed. Thus, even with a concerted effort to reduce the denominator, new desires will still appear and add to the denominator, making it constantly dynamic.

Now, this state of affairs may hold true for the majority of people who are solely focused on achieving the first three categories of goals. But what about the fourth goal—capturing the infinite? The goals in the first three categories are more turbulent, which makes the denominator constantly dynamic. However, because the fourth goal is a subtle, noble goal, the moment it has been set as one's primary goal in life, it has the effect of settling the denominator. Establishing this goal composes the mind, turning it into a dynamic constant. While the number of desires you entertain may still change, you will have more control over them, and will no longer suffer from agitation while you make an effort to pursue infinity. **In short, while capturing infinity is your primary goal, your personality will become a steady, dynamic constant.**

To illustrate this point, let's examine a hypothetical case using the happiness formula.

A case study of happiness

Imagine that you currently entertain 1,000 desires. This includes desires that you have already fulfilled and unfulfilled ones too. Also imagine that, based on your financial and other resources and as per your capacity, you have fulfilled 600 desires at this point in life. By using the happiness formula, we can determine that you are 60 percent happy.

Now, you wish to increase your happiness. Remember that there are two ways you can achieve this—by increasing the numerator and by decreasing the denominator. Let's assume that you focus your efforts on the latter. By making a careful effort to apply certain practices (which will be discussed in later chapters), you successfully eliminate 300 desires from the denominator. You now entertain only 700 desires in total.

At the same time, you used a portion of your resources to fulfill 50 of your remaining desires. You have now fulfilled 650 desires, out of a total of 700 desires entertained. Your happiness level has shot up to nearly 93 percent!

Let's say you continue your efforts, and eliminate 25 more desires, while also fulfilling another 20. With 670 desires fulfilled, out of a total of 675, your happiness level is now over 99 percent—an extremely high state of human refinement. But imagine that you go further and fulfill an additional four desires. This leaves a single desire unfulfilled, and your happiness at 99.9 percent.

What is that single unfulfilled desire? It is the desire that corresponds to the goal of capturing infinity.

Imagine your mental state at this point. When you entertain many desires, your flow of thoughts is tangled and turbulent; when you reduce them to just five unfulfilled desires, you achieve an extremely high state of human refinement, and your mental state becomes more composed. With a single desire remaining—and with that desire being the noblest of them all—your mental state will become more focused and composed than ever.

What will you be focused on? Remember that desires are streams of thought that flow toward a single object. With capturing infinity being your sole remaining desire, your thoughts stream smoothly, exclusively, and continuously in a direct line toward this desire.

Now, how long will this go on? What will it take to resolve this goal? With no other desires to focus on, your mind will be completely centered on this one desire from the moment you wake up to the moment you go to sleep. Thanks to our formula, we know that the net result changes either when the numerator increases or the denominator decreases. What will come into play here? Keep this question in mind as we briefly explore the method you will use to focus your attention on this one remaining desire.

How to capture infinity

With no other desires in your mind, you will sit all day with your flow of thoughts streaming directly toward this single desire. This is the state that is taught in the practice of meditation. Meditation will be discussed at great length later in this book, but it is necessary to touch on it now in order to describe the logic behind capturing infinity.

In the context of this example, your meditation will consist of sitting with one stream of thought toward one object—capturing infinity (consider infinity as a tangible object for now). **Meditation, in this case, is the capacity to direct the mind to a single flow of thought toward infinity, to the exclusion of all others, under the continuous observation of the intellect.**

Take a moment to consider how that definition of meditation fits in with your existing sense of the practice.

It's common for the practice of meditation to focus on clearing the mind, temporarily distancing oneself from desire. But, in your own experience, how long can that state actually be sustained? Typically for no more than a few seconds, or up to a few minutes.

This is because most people practicing meditation still entertain many desires. This makes it very difficult to distance oneself from the turbulence of the flow of thought and focus exclusively on a single object. Modern meditation practices, many of which are businesses, gloss over this fact. This is why many people approaching meditation experience mediocre results and eventually give it up—leading many people who begin by seeking *meditation* to wind up seeking *medication*!

But what happens when a person practicing meditation has only a single desire left—the desire to capture infinity?

With only a single desire remaining, you will be able to direct and sustain your flow of thought in a single stream toward it. This will enable you to practice meditation for extended periods of time. When your desires are reduced to the bare minimum, you essentially become *meditative*; you don't need to put any effort into meditating, because you don't have other desires to distract your flow of thought. You then spend your days directing your flow of thought into an uninterrupted stream toward your one remaining desire. What is the result of this? Let's draw a comparison from the world of engineering.

When testing the tensile strength of rods or malleable wire, researchers apply steady force to both ends of the object and begin to stretch it. As this happens, the diameter at the center gradually attenuates, until the material reaches its breaking point and snaps.

This same logic can be applied to the mind of a person in pursuit of infinity. During your meditation, your flow of thought will be stretched as it is directed toward infinity. If this effort is unbroken and sustained over sufficient time, it will eventually reach its limit and snap.

At this moment, your flow of thought ceases—and, in doing so, wipes out all desires, *effectively rendering the denominator in our equation nil.*

Returning to our previous example, you will now experience 674 desires (fulfilled) divided by 0 (having exhausted your last desire), which will yield ***infinite happiness***. In mathematics, dividing a number by 0 tends to infinity, an *undefined* result. This is another way to describe infinity, which is achieved when all your desires have been reduced to zero. You will be catapulted to an undefined quantum of happiness, and your personality will reach a state of infinity. Easier said than done! This is the logic behind capturing infinity.

Preparing to capture infinity

If it seems like we moved rather quickly through the central theme of this book, keep in mind that the purpose of this chapter was only to derive the *logic behind* capturing infinity. The remainder of this book is devoted to presenting in-depth lessons that prepare you for actually attempting to pursue and capture infinity. But in order to contextualize those lessons, it was necessary to present the logic behind this noble, rarefied goal.

The key takeaway here should be that your ability to capture infinity depends on your ability to eliminate all other desires besides that one.

Until capturing infinity is your sole desire, your attempts to achieve it will fail. The time it will take you to reach that state depends on how many desires you currently have, and how effectively you are able to both fulfill and eliminate them.

Once again, the main bottleneck here is the number of desires you currently entertain, and the rate at which you are able to reduce your unfulfilled desires. The outcome is entirely in your hands. It depends on your commitment to strengthening your intellect and controlling your mind through systematic methods and sustained, consistent effort. In the coming chapters, you will be presented with methods that will allow you to systematically eliminate and intelligently fulfill your desires. This will help get you closer to that state where capturing infinity is your sole desire. There, you will be one step away from reducing your desire to zero and finally capturing infinity.

In addition to covering these specific methods, future chapters will also discuss some of the questions that are likely already swirling around your head—questions that have to do with the creation of the universe, the Big Bang's prior mysteries, the meaning of capturing infinity, and what comes after this has occurred.

For example, we have established that the entire subtle body—mind, intellect, memory, and ego—is composed of streams of thought. When the stream of thought snaps, what happens to the subtle body? Do you lose your ego? Do you lose your individuality? Will you be more alive than ever, or will you drop dead? Will you escape the cycle of birth and death? These questions will be addressed in detail.

We will also spend some time discussing the scientific ramifications of capturing infinity. What does this mean for our understanding of the Big Bang, the universe, the multiverse, parallel universes, singularity, consciousness, causation, and creation? And how do these concepts relate to infinity itself? All of these interrelated concepts will be addressed in detail as well.

Though this book is not intended as a religious text, it should also be noted that, based on the nature of infinity, if such a thing as God exists, then God must be infinite. And because there cannot be more than one infinity, once one becomes one with infinity, one has essentially become one with that infinite God.

Because we have not gone into detail yet, it is natural for you to be wondering whether capturing infinity is possible at all. Can a human really entertain zero desires? Can one really eliminate all thought? Remember that the purpose of this chapter was only to introduce you to the logic behind capturing infinity, not to demonstrate how it is done. For now, just try to accept that the premise is possible, and that, once you have succeeded in eliminating all thoughts, you will be catapulted to infinity.

In the next chapter, we will begin the systematic presentation of the information, methods, and rationale that will enable you to pursue this noble goal. Each chapter is an essential part of your preparation, and as such, requires your undivided attention. If you make that commitment, you will set yourself up to achieve the rarest thing imaginable.

Chapter 6

Human Composition: Gross Physical Body

As we begin our systematic study of the nature of infinity and how it can be captured, we'll start with negations rather than affirmations. By analyzing what is *not* infinity, and continuing through a series of negations, we'll gradually develop a clearer understanding of what infinity is. The first subject we will cover in this portion of our study is human composition.

The human composition comprises 11 divisions, which are sub-divided and briefly explained below.

The first division contains the three classes of the body: the gross body, which we have briefly touched on already and will be the focus of this chapter; **the subtle body**, which we have discussed in some detail but will cover in much greater depth over the next three chapters; and **the causal body**, which we will spend two chapters discussing. **Together, these three bodies represent the material aspect of human composition.**

The second division of human composition contains the five classes of function. These are the five functional layers of the human personality, which will be discussed in detail in later chapters. Finally, **the third division of human composition contains three states of human experiences that we alternate**

between over the course of our lives. These are the waking state, the dream state, and the deep sleep state, and they will be covered after the chapters on layers of personality.

By thoroughly and systematically defining and exploring each of these factors, we will lay the foundation for a thorough understanding of human composition from various dimensions.

Analyzing the gross body

Our study of the three classes of the human body begins with an analysis of the gross body. **This analysis will consist of four stages. We will first study its raw material, then its specific components, its function, and finally, its nature.** Imagine you were studying the engine of a car—how would that process unfold? You would first take in its basic materials, then begin to recognize its finer components. This would lead you to develop an understanding of its function—an understanding that would culminate in an appreciation of its engine. We will adopt this same incremental process in our present study and will apply it in future chapters to the other two classes.

The raw materials of the gross body

As stated previously, **the gross body is simply the tangible, perceptible, and physical body.** All life forms have a gross body—plants and animals included. The gross body of a given organism is what distinguishes it as that specific organism. Put another way, the gross body of a living person and a recently deceased person are the same; but once a body has decomposed, it is no longer recognizable as the gross body. As long as the body is recognizable as a given being, we can refer to it as the gross body of that being.

The gross body is made up of gross matter—observable, measurable atoms. These are the raw materials of the gross body. To study it on the atomic and subatomic level, however, would require more space than we can devote here. For the sake of our primary objective, we will focus our analysis on the states of matter the gross body is made of. This will allow us enough nuance to understand the premise without distracting from the broader purpose of the book.

The human gross body, then, comprises a combination of the various states of matter. From a young age, we are taught that there are three fundamental states of matter—gas, liquid, and solid. In reality, there are *five* fundamental states of matter, and these five states combine to form the gross body. Let us explore each of these five states in the context of human composition.

The first state of matter is the gaseous state, otherwise known as air. Air is essential for sustaining human life, and it is a part of the gross body. After death, air gradually leaves the body. But during life and also during the period immediately after death, when the gross body is still recognizable, air represents one of the raw materials of human composition.

The second state of matter is the liquid state, also known as water. About 60 percent of the human body is made up of water. Like air, gradually water too leaves the body after death. For as long as the gross body is recognizable, however, water is a key part of its composition.

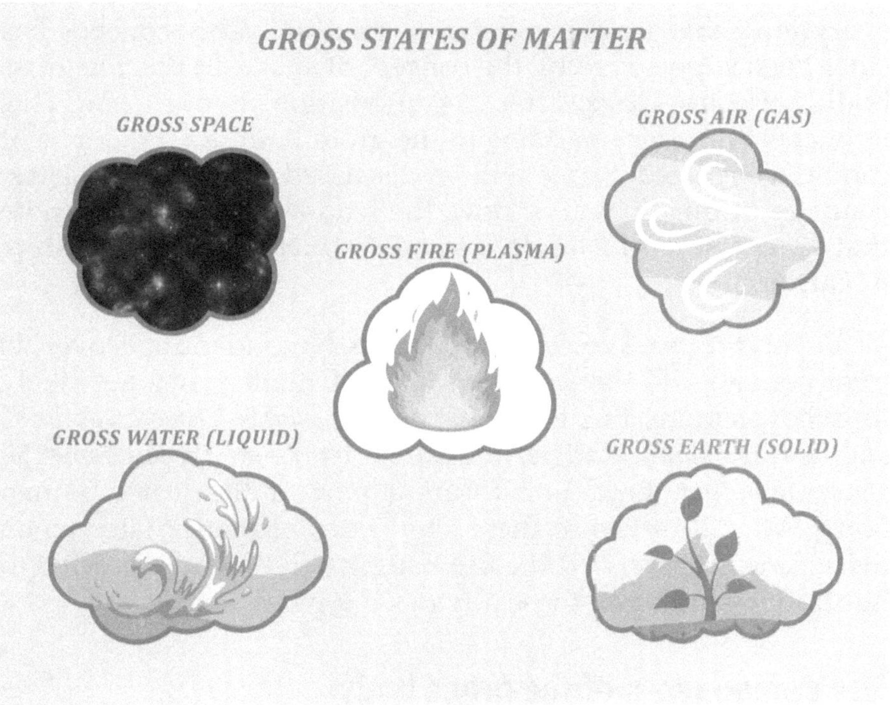

The third state of matter is the solid state. The structural components of the gross body consist of matter in the solid state—the flesh, bones, and other organs. This state corresponds to the classical element of the earth, and these structural components return to earth after death and decomposition.

Two additional states of matter contribute to the composition of the gross body. The first of these, **the fourth state of matter, is plasma, also known as fire.** We can see that this state of matter comprises part of the gross body from the fact that the human body maintains a temperature of roughly 98 degrees Fahrenheit throughout life—and that, after death, this temperature vanishes.

The fifth and final state of matter that comprises the gross body is space. In the past, scientists entertained the concept of *ether* as a medium that would allow light to travel, since everything that travels requires a medium. This concept has since been superseded by the concept of space. In life, the gross body occupies space; after decomposition, it does not. This indicates that space pertains to the gross body for as long as it can be recognized. Space will be discussed further in the latter chapters of this book. For now, the reader should simply note that ***space not only is matter, but a distinct, fundamental state of matter.***

Beyond these five states, scientists have identified over 20 intermediate and/or extreme states of matter, and are likely to uncover more. But these *five primary states—space, gas (air), plasma (fire), liquid (water), and solid (earth)—are the fundamental states of matter.* Each one forms a part of the gross human body. We will return to these five states of matter later when discussing creation and the Big Bang when they will be vital to our understanding of these advanced topics.

The components of the gross body

Now that we have established the raw materials of the human body, we can extend our study to the body's components. These components are the distinct anatomical parts of the body. Once again, these specific components are too numerous to discuss and analyze in this book. Interested readers may wish to study human anatomy independently. For our purposes here, let us simply name and briefly explore some of the body's primary physiological systems—specifically those that are involved in actively contributing to one's experience of life.

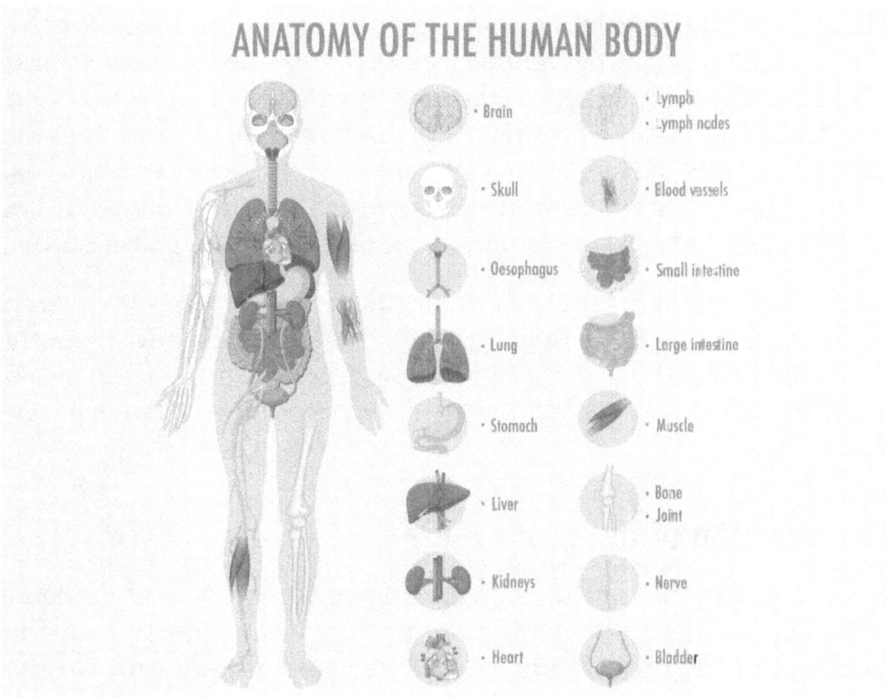

What does it mean for a system to actively contribute to one's experience of life? These are the systems that enable us to perceive stimuli originating from the broader world, react to them internally, and enact some external response. The primary gross organs involved in these processes are listed below.

The first of these are the gross organs associated with perception—the eyes (*sight*), ears (*hearing*), skin (*touch*), tongue (*taste*), and nose (*smell*). **Next are the organs and structures associated with performing actions**—the mouth (*speech*), the hands (*manipulation*), the legs (*locomotion*), the excretory organs (*excretion*), and the reproductive organs (*reproduction*).

Then there are the numerous involuntary organs and bodily systems that sustain life without our intervention. These include *respiratory* organs, like the lungs, trachea, bronchi, and alveoli; internal *excretory* organs like the liver, kidney, and bladder; *digestive* organs, like the esophagus, stomach, and intestines; *circulatory* organs, like the heart and blood vessels; and the organs of the *nervous* system, like the brain and nerves. Each of these organs and systems contributes in distinct ways to the sustained healthy functioning of the human gross body.

As we saw with the raw materials of the gross body, each of these systems is present in both living bodies and recently deceased ones. *While it may seem strange to keep bringing up death in this way, the rationale will become clearer in later chapters on the subtle body.*

The function of the gross body

Now that we have an understanding of the gross body's components, we are ready to explore its function. **Simply put, the function of the gross body is to serve as a housing unit for the subtle body, enabling it to act in this world.**

Because each subtle body is specific to a particular gross body, the type of gross body one acquires at birth affects the scope of actions available to one's subtle body. Because one's birth is determined by actions taken in past lives, it can be said that **the ability of the gross body to carry out its function is the product of one's actions in past lives.**

Recall the principle of invisible wealth discussed in the chapter on human goals. The invisible wealth accumulated in past lives determines one's birth in this life. This includes the gross body one acquires. It affects everything from body type, genetics, and physical health to the environment of one's birth. One can be born into a poor environment that will harm one's gross body, or a rich environment that will nourish it. The form of one's gross body is the cumulative result of both positive and negative actions done in the past.

Now, what does this have to do with the function of the gross body? Consider the way one's gross body affects the actions available to a person. Generally speaking, a person born with a weak gross body in an impoverished environment will take very different actions in life than a person born with a strong gross body in a privileged, comfortable environment. Of course, one's actions are not predetermined, and people constantly take actions that deviate from what one might expect. Even so, the form one's gross body takes *does* have a significant bearing on the way it functions.

This same principle can also be extended into the future. While your past actions determine the form of your gross body in this life, your actions in this life will affect its form in the next life. Thus, this function exists on a continuum, affected by past actions and effecting future ones.

In principle, this follows the same logic as the positive and negative resultant forces that we studied in physics. This topic will be revisited in detail in the chapters on causation, but it is useful to bring it up now because it aptly describes the relationship between our gross body's form and the actions we take.

Before moving on to the nature of the gross body, let's observe one final point about the gross body's function. Not only does it allow the subtle body to act, but also allows it to experience pleasure and pain. Just as the form your gross body assumes is determined by your past and present actions, the pleasure and pain you experience are determined by your actions as well. **In other words, the function of the gross body is to serve as the medium for your subtle body to both exert its actions and experience the consequences of them.**

The nature of the gross body

Having established the materials, components, and function of the gross body, we are now ready to explore its nature, which consists of four main aspects.

The first aspect of the nature of the gross body is cyclical modification. Over the course of one's life cycle, one's gross body goes through six distinct modifications.

The first of these modifications is the transition from potential existence to growth in the womb—one of the most complex and wonderful phenomena in the entire observable universe.

The second modification is birth—the transition from life inside the womb to life in the world.

The third modification is early growth—the period of relatively rapid growth from infancy to the age of roughly 20, when the body takes on its adult form.

The fourth modification is mature growth—the period during which external growth stagnates, but the body continues to age and mature.

The fifth modification is physical degeneration—the period during which the body begins to slow down, and its systems begin to weaken and fail. This process continues up until the sixth modification, the final, inevitable modification that awaits every single human being—death. After this, the cycle of modifications begins again, and so on. **This cycle of modification represents the first nature of the gross body.**

These modifications unfold steadily and continuously, with one flowing seamlessly into the next. Each and every day, countless changes occur within the body. Cells die and are born, and all the while the body progresses from one modification to the next in a state of continuous change. **This is the second nature of the gross body—continuous change.**

The third nature of the gross body is that its existence is finite. The current average worldwide life expectancy is just over 72 years; the oldest person who ever lived made it to just over 122. Of course, the body can expire at any time—even before birth. But it is a certainty that death *will* come for every gross body, which makes this attribute one of its fundamental natures.

The fourth nature of the gross body is that it is tangible and perceptible, both to oneself and to others. After all, it is composed of gross matter, which is inherently tangible and perceptible. Hence, it is a fundamental nature of the gross body.

This concludes our overview of the gross body and sets up our study of the subtle body, which will begin in the next chapter. As part of that study, we'll first take a moment to clearly define *gross* and *subtle* in the context of one another. The term *subtle* has been used extensively thus far but has not been defined completely. The term Gross and Subtle shall be explained in the next chapter.

Chapter 7

Human Composition: Subtle Body - Part 1

The previous chapter introduced the concept that the human composition consists of 11 factors spread across three categories—body, personality, and states of experience. Within the first category, there are three classes of body—the gross body, the subtle body, and the causal body. We also conducted a brief study of the gross body. Now, we are ready to move on to the subtle body.

First, let us take a moment to properly explain the terms *gross* and *subtle*. These terms are relative, existing within a hierarchy of subtlety that identifies the nature of one object based on its relationship to another. How is this determined? Let's examine that question before moving on to the broader discussion.

The hierarchy of subtlety

According to the hierarchy of subtlety, **Object A is subtler than Object B only when A controls B *and* also pervades B**. These two factors—control and pervasiveness—determine the order of subtlety.

For example, we can say that the mind is subtler than the gross body because the mind both controls the gross body (by directing it to act) and pervades it (by existing within it and beyond).

Of the two criteria, the concept of control is easier to understand. The mind controls the gross body, and not the other way around. Can a dead body act? No. Therefore action requires the direction of the mind.

The concept of pervasiveness is a little trickier. It can be understood by asking whether an object can or can't go everywhere another object does. For example, everywhere the gross body goes, the mind goes, too; but everywhere the mind goes, the gross body can't always follow. When you think of the sun, your body can't physically go there with it. So, we can see that the mind pervades the gross body, but the gross body does not pervade the mind. Thus, since we have seen that the mind both controls and pervades the gross body, we can say that it is subtler than the gross body.

Likewise, the intellect is subtler than the mind, since the intellect both controls the mind and pervades it. Wherever the mind goes, the intellect can also go; however, the unique power of the intellect is that, whenever the mind gets stuck, the intellect can pierce through and open new avenues of thought. So, the intellect can be said to be subtler than the mind. This also implies that the mind is grosser than the intellect. This is the nature of the hierarchy of subtlety, in which the terms *gross* and *subtle* are relative.

How did we end up with the terms gross body and subtle body? Simply put, we call the gross body *gross* because it is not subtler than anything. We call the subtle body *subtle* because it wholly pervades and controls the gross body. Even though there are gradations of subtlety and grossness within the subtle body, as a whole it is subtler than the gross body.

Going forward, the reader should understand that the terms *gross* and *subtle* invoke the concepts of control and pervasiveness. This knowledge will be important as we conduct our analysis of the subtle body.

Subtle body and subtle matter

We will consider the subtle body in the same four stages we used during our analysis of the gross body. What is the raw material of the subtle body? What are its components? What is its function? And lastly, what is the nature of the subtle body?

Till now, when we referred to the subtle body, we focused on four distinct components—mind, intellect, memory, and ego. There are actually an additional 16 components, making a total of 20 components within the subtle body. As with the four components we've already discussed, the fundamental building block of these other 16 components is thought. *Thought* is the raw material of the subtle body. In the next chapter, we will identify each of these components; here, we will concentrate on developing a clear understanding of what exactly this raw material is.

Exploring the nature of subtle matter

Once again, **thought is the raw material of the subtle body**. But that statement is not enough, because it begs the question: What are thoughts made of?

In the previous chapter, we concluded that the gross body is made of gross matter. We also identified the five fundamental states of gross matter—space, fire, air, water, and earth.

Just as the gross body is composed of gross matter, *the subtle body is composed of subtle matter.* Because we know that thoughts are the building blocks of the subtle body, we can extend this statement to reveal that **thoughts are composed of subtle matter.** So what is subtle matter?

We will discuss this ***subtle matter*** extensively in the chapters on creation, where we will explore its origins. But we will briefly define this complex term now because it is necessary to understand its composition.

Just as people wind up with a wide variety of gross bodies from the same building blocks, they also wind up with various subtle bodies based on the same raw material—subtle matter. The mind and intellect of each individual are unique because they (like the gross body) are the cumulative result of an individual's past actions.

To draw a comparison, while walking through a city you will encounter all kinds of buildings, from modest homes to large apartments and great skyscrapers. These buildings differ in both appearance and purpose, but they are all made up of brick, cement, glass, and steel—the same standard raw materials used for construction. Their differences arise from many sources—their blueprints, their intended purpose, the skill of the builder, and the availability of materials. In other words, each building is the cumulative result of all decisions that went into its planning, which determine its final, unique form.

The same thing occurs during the formation of the gross and subtle bodies. We already saw how this affects the gross body. For the subtle body, one's past actions affect one's mental temperament, the strength of intellect, and every other component of the subtle body. And all of these components are composed of subtle matter.

Revealing the nature of subtle matter

Let's move on to the definition of subtle matter. This will be described in much greater detail when we arrive at the subject of the Big Bang and creation.

First, let's invoke a lesson from the field of quantum mechanics. Those who have studied the field will likely be familiar with the Pauli exclusion principle, which holds that no two electrons can occupy the same quantum state simultaneously. That is, within atoms, electrons are arranged into distinct orbitals that prevent them from occupying the same place at the same time. The gross matter that is a single electron occupies space that is all its own, to the exclusion of any other gross matter.

To translate this into a more tangible example, if somebody is sitting in a chair, another person cannot come and sit in the same chair, nor can they pass through the person or occupy the same space as them in any way. This, in essence, describes the Pauli exclusion principle. Now, let's see how this concept is relevant to subtle matter.

Recall that gross matter is composed of the combinations of five fundamental states of matter—space, fire, air, water, and earth. **Subtle matter, on the other hand, is composed of combinations of the *five pure states of matter*.** Now, what is meant by pure states of matter? Let's consider an example to help unravel this point.

First, consider a common state of gross matter—say, water. Now consider this water in its *pure state of matter*. In order for this water to truly exist in its pure state, it must consist of *only* liquid, to the complete exclusion of the other remaining four states of matter. It must not contain earth, air, fire, and space.

Water, in its gross state, generally includes some amount of the other four states of matter. All gross water has some temperature—that is, some fire, some traces of minerals—that is, some earth. It also contains some air, and most problematic of all, *all gross water necessarily contains some space.*

How, then, can it be possible for matter *(water)* to exist in a pure state? How can all other states of matter be completely excluded?

The answer is that **the pure state of any form of matter *can only exist in thought*.** Put another way, thought is the only environment in which all other states of matter can be excluded, leaving only one type of matter in its pure state.

It's important to note that this doesn't mean that you have simply 'imagined' pure water or as if that were sufficient to purify a state of matter. Instead, the pure states of each type of matter can only exist in the form of thought. *When the five fundamental pure states of matter combine, they take the form of thoughts.*

An example may help to clarify this point. Think of a tangible object—like a car, for example. By entertaining this thought, you have not realized a 'pure state of car,' because there is no such thing. A car is a complex object that is composed of varying amounts of these five gross states of matter; it can't magically attain a pure state of its own just by becoming a thought. Nonetheless, you are thinking of a car—so what *is* that thought?

The answer is that it is a combination of the pure states of the five forms of matter, which exist in thought form and have combined to create the thought of a car. In the same way that a physical car is made of a combination of gross matter, the thought of a car is made of a combination of pure—or, rather, *subtle* matter.

As further proof, let's return to the example of the pure state of water. What happens if you attempt to remove the pure state of water from that thought? No thought will remain. The thought of water will be eliminated. This corroborates the idea that thoughts are composed of subtle matter.

As we close this section, recall the Pauli exclusion principle, which contributes to the stability of gross matter. This principle fails at the level of subtle matter because multiple pure states of matter can combine to form a single thought. And yet it does not create instability, because it is in their nature.

Matter, thought, and consciousness

So, thoughts are composed of subtle matter and combinations of pure states of matter. But this explanation seems to introduce a new wrinkle—**if the subtle body is composed of thoughts, and thoughts are composed of matter, then how can the subtle body also be conscious?** In order to respond to this question, we must once again pull in ideas that will be discussed at greater length later in this book. A later chapter will be devoted to consciousness (and its relationship to the Big Bang, creation, and infinity), but it's necessary to give some attention to this topic now to better understand the subtle body.

Whether it's gross or subtle, all matter obeys certain fundamental rules. One of these is that matter is intrinsically inert and insentient. An inert object is something that cannot be conscious by itself, and such an object also cannot generate consciousness. And yet, the subtle body *does* possess consciousness. How does this happen?

The subtle body is sentient. Yet matter is intrinsically inert, and thus *should not* be sentient; no matter whether it's gross or subtle, matter should neither possess nor produce consciousness.

Though thoughts, subtle matter, and the subtle body are material in nature, they borrow their consciousness from what is known as the *infinite consciousness principle.*

Subtle matter is bathed in the medium of infinite consciousness, which allows the subtle body to become sentient. This principle is too nuanced and hence will be discussed in a later chapter. For now, simply posit the existence of an infinite, all-pervading, consciousness from which matter borrows consciousness and draws sentience.

If matter is intrinsically inert and cannot be sentient or generate consciousness by itself, then it must also be the case, that consciousness—is and must be non-material in nature. We will elaborate on the nature of consciousness later when we prove that it is eternal, homogeneous, all-pervading, independent, non-dualistic, and infinite.

Our extremely brief introduction to this complex point has no doubt left you with many questions about how consciousness is 'borrowed.' For example, why aren't *all* inert objects borrowing consciousness and gaining sentience? Wood, sand, metals—why don't they act like 'Transformers' and become conscious like human beings? This is a valid question and one that we will answer in due course.

This discussion has brought us a little out of the way of our analysis of the raw materials of the subtle body. To recap, thoughts are the building blocks of the subtle body; subtle matter is the raw material of thought and is composed of combinations of pure states of matter. Subtle matter borrows consciousness from an all-pervading infinite medium of consciousness.

Before concluding this chapter and moving on to discuss the components of the subtle body, let's take a moment to examine a few last points on the nature of thought and concentration. This will help to further refine our understanding of the raw material of the subtle body.

The speed of thought

Let's recall the first chapter, where we first touched on the fact that nothing can travel faster than the speed of light—although, in our expanding universe, space can *appear* to be traveling faster than light.

Knowing what we know about the nature of thought, though, can't it be said that thought can travel faster than both light and space?

The moment you think of the sun, your thoughts can alight on to the sun; the moment you think of some distant galaxy, your thoughts can travel that far! Consider the latest discoveries of the JWST, revealing secrets at the so-called edge of the universe. The moment these things are discovered, our thoughts can go there instantaneously. No matter where we are, within nanoseconds our thoughts can travel over the vastest distances imaginable in the observable universe.

This seems to suggest a paradox—after all, thought is made of matter and yet it is capable of traveling faster than the speed of light. This seems to beg new thought experiments, like those exploring superposition or quantum entanglement. Such questions examine the puzzle of how matter behaves under varying states of observation—behaving like a particle when observed yet like waves when not.

This invokes the interconvertibility of matter and energy, which points to another aspect of the nature of thought. Much like how matter gets converted into energy to generate electricity for your home, subtle matter is converted into thought energy during the process of thinking. Since the mind is constantly swirling with countless desires, which are composed of thoughts, one is continuously burning subtle matter to generate the thought energy required to handle this load.

This means that the way you generate and allocate thought energy is very important. Specifically, it's important to avoid wasting thought energy on useless activities, like excessive worrying about unknowns or reflection on past events or things that are beyond your control.

Burning subtle matter on useless thought energy is like leaving all the lights on in your house 24 hours a day—it's a waste. We all want to save money in the material world and yet are comfortable wasting the equivalent in terms of subtle resources by entertaining anxieties about the outcome of future events i.e. generating countless thoughts. In the future, one will experience only a single effect of the cumulative causes leading up to it. Yet, that person will nonetheless imagine countless imaginary outcomes, which is a sheer waste of energy.

If only we were able to save all that wasted mental energy and use it on our current ongoing activities instead—that is, to truly *concentrate*—we could become much more deliberate and efficient in our actions. Recall that concentration is the capacity of the intellect to keep the mind engaged in its present activity, not permitting it to generate thoughts about the past or the future.

Keep these two last observations in mind: despite being matter, thought can travel faster than the speed of light; and thought energy can be wasted through unnecessary worries, excitements, and anxieties. These ideas will bring valuable nuance to this concept and will be helpful as we move to the next stage of our analysis of the subtle body.

Chapter 8

Human Composition: Subtle Body - Part II

After discussing the raw material of the subtle body, we are ready to move on to our analysis of its 20 components. However, for convenience, we are going to make a slight deviation from the four-fold analytical approach we have already proposed. This is because elaborating on the details of 20 components would be cumbersome and confusing. Instead, as we identify each component of the subtle body, we will also describe its function. This will be more orderly and efficient and will help us continue our discussion about the nature of the subtle body in the following chapter.

We compared the gross body to a housing unit that hosts the subtle body earlier. The components of the subtle body, then, can be compared to the sophisticated equipment of a modern home. If the gross body is the equivalent of bricks and concrete, then the components of the subtle body represent the appliances and electronics that enable functional and efficient life within it.

The 20 components of the subtle body are split into four categories of five components each. These categories are defined by the functions of their components. This is another reason why we should discuss each component's function as we introduce and define it. That said, let's begin our analysis of the subtle body's components, starting with the first category—the five power centers of perception.

The five power centers of perception

The five power centers of perception, also known as the five power centers of knowledge, enable humans **to register sensory stimuli originating from the outside world.** When we discussed the gross body, we mentioned the five sense organs—ears, skin, eyes, tongue, and nose. They are a part of the gross body, meaning they are present in both a living and a dead body. However, the latter does not possess the capacity for perception; therefore, the power of sensory perception is something that exists separately from the sense organs that enable it.

These components are intermediaries within the subtle body that enable coordination between the gross sense organ and the mind—coordination that takes the form of thought. In this way, the power centers of perception act as bridges between the mind and the gross sense organs.

Why do we call these components the power centers of both perception and knowledge? It's because the sensory information entering these centers becomes the raw data our mind and intellect use to build intelligence and make decisions. All sensory perceptions that enter our gross sensory organs flow through the power centers of perception toward the mind and the intellect. All these stimuli represent new information for the mind and the intellect to use, which is why perception and knowledge are combined in them.

The powers of perception can be understood as the subtle equivalents of their gross counterparts. The perceptive power of the ear, for example, can be referred to as the subtle ear—as opposed to the gross ear, the actual organ that takes in auditory stimuli. Similarly, we have the subtle skin, the subtle eye, the subtle tongue, and the subtle nose.

The strength of these power centers varies from one person to the next. As an extreme example, consider the title character of *The Bourne Identity*, who has received special training that has made all of his senses extremely sharp. Contrast this with something you might have experienced in your own life. Have you ever approached one of your friends and stood right in front of them, even speaking directly to them, and yet it has taken them a few moments to register that you are there? So, there are gradations when it comes to the power of these subtle perceptive components.

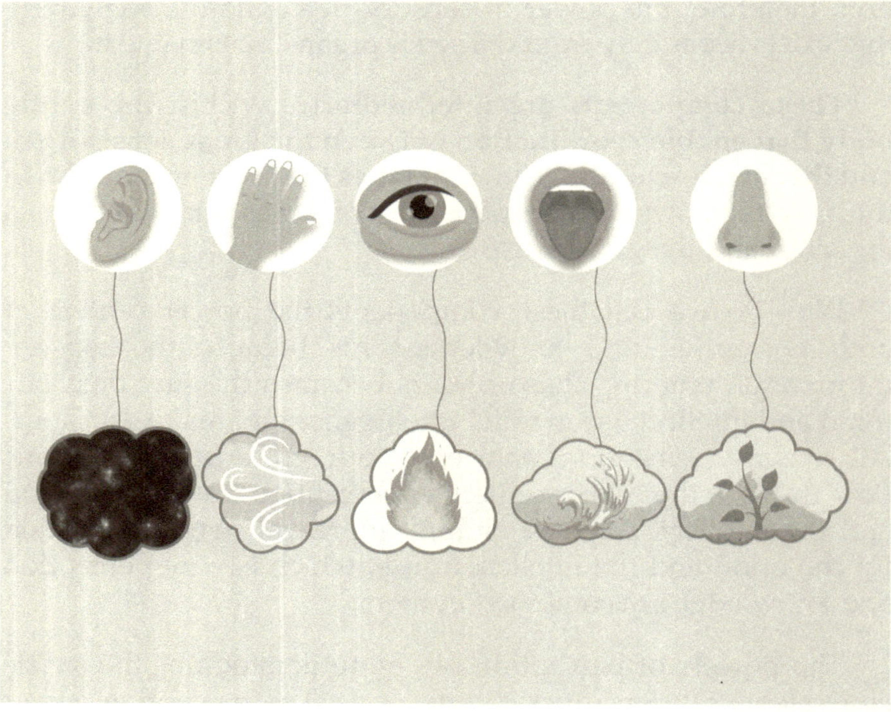

It should also be noted that each of these power centers is confined to its own faculty. They cannot be interchanged. For instance, the subtle eye cannot replace or fulfill the function of the subtle ear, or vice versa. Each one is exclusively suited to the performance of its specific function. This is an important limitation to keep in mind when contemplating the function of these components.

Before we move on to discuss the next category of subtle components, take a moment to look at the image. It presents the five sense organs of the five power centers of perception, along with the five states of matter. Each of these power centers is associated with one of the states of matter. Specifically, each power center is paired with the most rarified state of matter that it is capable of perceiving. What does this mean? Let's get to the specifics of each component.

We'll begin with space. One cannot feel, see, taste, or smell space—but one can hear it. Thus, space is one of the most rarified states of matter, and hearing is associated with it because it is the only sense capable of perceiving it.

Next is air or gas. Air cannot be seen, tasted, or smelled—but it can be felt and heard. Thus, the skin is associated with air, because the power center of touch is capable of perceiving it.

By this same logic, the eyes are associated with fire, or plasma, because fire can be seen, felt, and heard, but not tasted or smelled. The tongue is associated with water or liquid, because water can be tasted, seen, felt, and heard, but not smelled. And, finally, the nose is associated with earth, or solid matter, because, while solid matter can be perceived by every sense, it is the only state of matter that can be perceived by smell.

This information will become particularly relevant when we come to the chapter on the Big Bang and creation. For now, simply note this point on the relationships between the states of matter and the subtle sensory components.

The five power centers of action

When the gross sensory organs receive external stimuli, they pass them through the power centers of perception to be processed by the mind and the intellect. Once the stimuli have been processed and the intellect has determined an appropriate response, an action is triggered to carry out the response in the physical world. **These responses are enacted by the five power centers of action, which enable the physical manifestation of thoughts. They are the next category of components of the subtle body.**

These components correspond to the gross organs and structures associated with action, which were described in the chapter on the gross body—speech, manipulation, locomotion, excretion, and reproduction.

Of these, the most powerful is the power center of speech. In fact, after the intellect, speech is the most powerful subtle organ in the body. Why? The human capacity for speech and communication is responsible for all the progress and learning our species has experienced. Speech is enduring. There is a saying in Tamil that roughly translates to, *"The scar on your skin caused by a burn will fade away in time, but when you burn your fellow beings with hurtful words, those scars will never heal."* This sentiment indicates the power of speech.

The next power center is manipulation. In the gross body, this is centered on the hands. It enables the human capacity to build, create, and otherwise manipulate physical matter. After that is the power center of locomotion, centered on the legs. It allows us to move, travel, and run toward or away from danger.

Next is the power center of evacuation, centered on the excretory organs and systems that enable the elimination of bodily waste—an essential function of healthy human activity.

And finally, the power center of reproduction, centered on the reproductive organs. It is responsible for the human capacity to create progeny—a critical power, without which our species would cease to exist.

As with the five power centers of perception, these five power centers of action are centered on gross physical organs. They are not the organs themselves, but rather the subtle counterparts of the gross structures that enable them to interact with the world.

With that, we are ready to move on to the next category of subtle components—the five power centers of physiological function.

The five power centers of physiological function

The five power centers of physiological function are responsible for enabling the internal processes that sustain **a healthy life.** They ensure that the gross organs of perception, the gross organs of action, and the entire gross body are healthy and nourished so that they remain healthy.

The first of these centers of physiological function is the respiratory system, which enables inhalation and exhalation. It powers the human capacity to take in air through the nose and mouth and to direct it through the trachea and bronchi to the lungs, from where oxygen passes into the blood. As with all the other power centers, this power center is not identical to these gross physical structures but is a subtle version of them that enables their complex function.

Next is the power center of the digestive system. It enables food to be broken down into usable resources and made available to the entire body as nutritive fuel. This process begins with mastication and continues with digestion and nutrient extraction. It also involves the separation of waste and toxins from usable nutrients. It is closely connected with the power center of excretion, which we will touch on momentarily. In fact, all of the power centers of physiological function are closely interrelated, directing the passage of gross matter from one system to another to ensure the body's sustained health.

The third power center is that of the circulatory system, which enables the circulation of vital nutrients throughout the entire gross body. It also transports waste material to the excretory system to be removed. The power center of the excretory system then enables the separation of all gross material waste from the other physiological systems via the organs of excretion.

Finally, there is one last physiological system that is often overlooked but is nonetheless extremely important. This is the body's emergency shutdown system. It is governed by the power center of reversal. To understand the purpose and function of this power center, consider the principle of denaturation. Many substances, when heated over a period of time, liquefy. For example, a metal that is heated for long enough at high temperatures will melt. Other substances, such as eggs and certain proteins, denature and become solid when heated. This is the principle of denaturation, and it indicates how certain substances can behave differently under the same circumstances.

In the human body, a similar principle holds true. When the body takes in food, for example, the digestive process is normal. But when it takes in toxins or certain types of food—say, undercooked or contaminated food—that process does not function properly. In such cases, the power center of reversal steps in and directs the gross body to eject the improper substance. In this scenario, that would be through vomiting.

This power center functions in other ways, as well. For example, if dust or sand enters the eyes, it immediately produces tears in order to clear it away. If a foreign object enters the nose, mucus is secreted in an attempt to flush it out, and the sneezing response may be triggered. This power center is not active at all times, as the other four are; it only acts on an emergency basis. This is why it is also known as the power center of emergency shutdown.

This power center fulfills a critical function at the time of death. After death, all the physiological systems slow down and then stop altogether. This process is facilitated by the power center of reversal, which helps to extricate the subtle body out of the gross body.

This claim will be substantiated with additional proof when we get to the subjects of reincarnation and causation. For now, let us continue to the final five components of the subtle body.

The final five components – The inner personality

The final five components of the subtle body make up what we have previously referred to *as* **the subtle body—that is, the mind, intellect, memory, and ego. This group is more accurately known as the inner personality.** While we have referred to the intellect as a single entity up to now, there are actually two kinds of intellect—gross and subtle—which is how we have five components in this category. In the same way that the intellect is subtler than the mind, within the intellect there is one portion that is subtler than the other. We will discuss this distinction momentarily; first, let us begin with the subtle body's 16th component—the mind.

The 15 power centers are the bridge to the mind

We have already discussed the function of the mind throughout this book. Suffice it to say that the mind is the seat of all desires, emotions, dilemmas, confusions, and misunderstandings—all these things are present in the form of thoughts.

The mind is the entryway to the inner personality. The 15 power centers we have discussed throughout this chapter are the bridge between the gross body (which takes in stimuli from the outside world and effectuates physical actions) and the inner personality (which possesses the faculty of examining those stimuli and directing one's physical responses to them).

How does this work? Let's start with the five power centers of perception. External stimuli enter the gross sensory organs and pass through the relevant sensory power center to the mind. In this way, the power centers of perception are the bridge between the stimuli and the mind.

Once the stimuli have been registered in the mind and the intellect has determined what to do about them, it passes those decisions back through the mind to the five power centers of action. In this way, the power centers of action are the bridge between the mind and the actions taken by the gross body.

Finally, for the body to continue functioning properly, the mind also coordinates with the physiological systems via the power centers of physiological function. In this way, these power centers form the bridge between the mind and a healthy, functioning body.

Taken together, the mind functions as an intermediary between the inner personality and the subtle body's centers of power.

Gross intellect and subtle intellect

After the mind comes the intellect—or, rather, the *intellects,* since there are actually two types. These are the gross and the subtle intellects. In both cases, the intellect enables the same faculty—the capacity to form discrete thoughts, reason, judge, discriminate, validate, etc. Let's explore the distinction between the two types of intellect, along with the real-world outcomes for individuals with varying strengths of intellect.

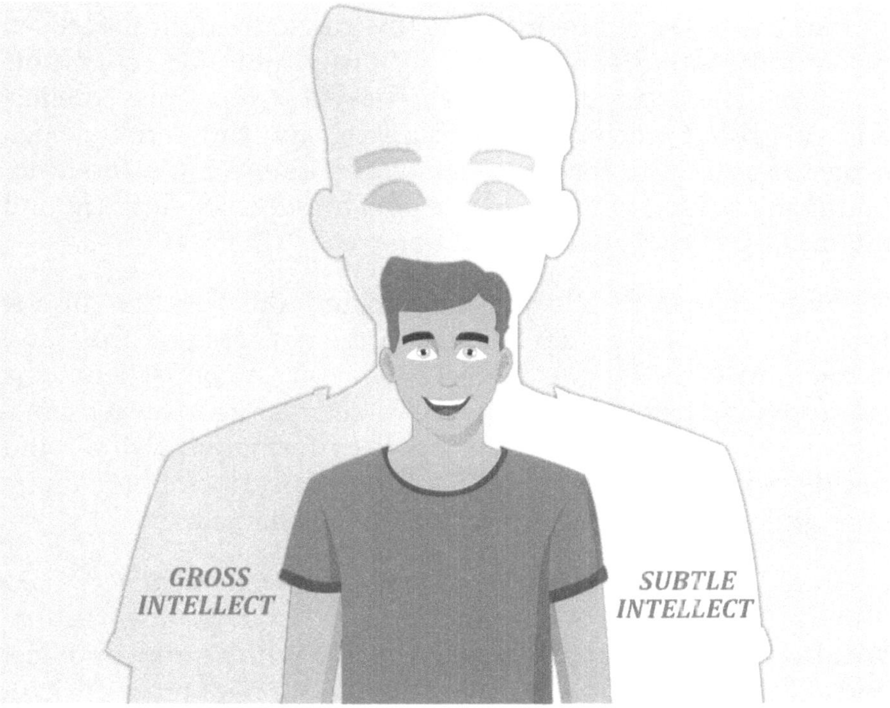

What distinguishes the gross from the subtle intellect? **The gross intellect is the capacity for forming thoughts that deal with the observable universe.** Any thought that has to do with the observable universe is performed by the gross intellect. This includes all the objective analysis and all scientific experimentation.

The subtle intellect, on the other hand, is the capacity for forming thoughts that go beyond the observable universe. These are thoughts that entertain the existence of something beyond the universe, the possibility of infinity, and the possibility of *capturing* infinity. Those rare 10 million thinkers engaged in the pursuit of capturing infinity use their subtle intellect when they entertain these thoughts, beyond the observable universe.

All intellect is the faculty of forming single, discrete thoughts. In the case of the subtle intellect, this single thought is focused on infinity, on distinguishing infinity from the observable universe. There may be those who have a strong gross intellect and yet cannot project the possibility of capturing infinity. Such a person is said to have a weak subtle intellect. On the other hand, any person who is capable of contemplating infinity, and attempting to capture it, is said to have a strong one.

Now, in order to better understand the function of the intellect, let's take a step back from the concepts of gross and subtle and consider the intellect as a single component. The reader will note that the intellect doesn't operate in a vacuum—it requires inputs in the form of data/information to infer and decide. So, the function of the intellect depends not only on its strength but also on the inputs that are available to it.

Recall the distinction we made between intellect and intelligence—the latter referring to a pool of data/information. We also indicated that a genius is a person with a strong intellect and a high degree of intelligence. In the past, it was more difficult to access and acquire information, so geniuses were rarer. Now, with the internet and other modern resources that have made information available, gaining intelligence is much easier, and being a genius is more common. In essence, one only needs to possess a strong intellect in this era in order to become a genius.

What are the outcomes from other combinations of strength of intellect and degree of intelligence? How does the intellect function under those circumstances? Let's look at a few examples in order to understand.

Consider an individual with a strong intellect but who, for whatever reason, lacks the resources to acquire much intelligence. This person will still make good decisions, although it may take longer to do so because acquiring the necessary data will take longer. But the outcomes will still be positive.

Now, consider an individual with a weak intellect but because of the resources at his/her disposal, has strong intelligence and ample data. This person will also make good decisions, although it will be more a result of intelligence than intellect. Nonetheless, this combination results in positive outcomes as well.

Consider now the category of people who have negligible intellect. These people may have lazy, animalistic, or highly inert natures. For these people, it doesn't matter what degree of intelligence they are presented with—the outcome will always be negative. A negligible intellect always produces poor outcomes.

Finally, there is a rare fourth category—individuals with a strong intellect, but with access to the *wrong intelligence.* This combination will also produce poor outcomes. This generally happens when an individual trusts people to give good intelligence but receives the wrong kind. In such cases, it doesn't matter how strong one's intellect is. One could have the strongest intellect in the world and would yet make bad decisions.

All this goes to show that, while the intellect's function is to enable discrete thought, the outcome is also dependent (to an extent) on the intelligence available.

The memory

The next component of the subtle body is the memory, which records and stores all of one's experiences. **The human memory**

stores the experience from all five perceptions, all actions, and all thoughts and feelings—in short, everything.

This does not, however, mean that we can recall everything that is stored in our memory. Despite having a large enough capacity to store all experiences, our ability to *retrieve* specific memories depends on how our memories are stored.

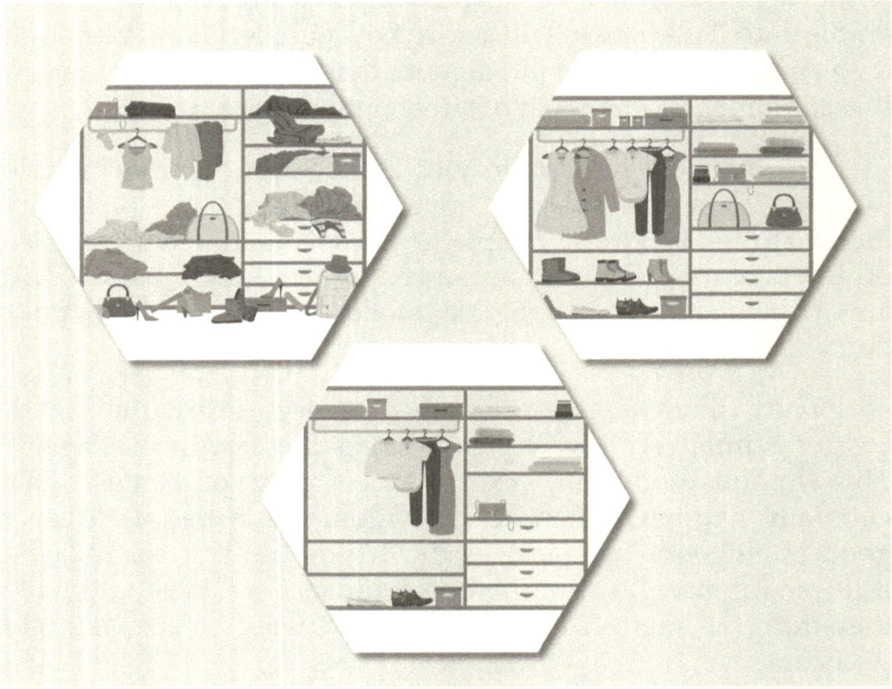

Consider how different people arrange their closets. Some people have very messy closets, with clothes strewn everywhere. It would be very difficult to find anything in such a closet. Another person's closet might be more orderly. This closet may contain exactly the same clothing as the other person's, but it will be easier to find things. Finally, consider a third person who has a minimalist lifestyle. That closet will also be orderly because of the fewer things in it. It will be even easier to find things as there will be fewer things to look through.

Extending the example further, the way someone arranges their closet reflects their general approach to life. People with a more chaotic approach may just throw things in their closets without much care for where they land. Other people may live a more standardized lifestyle; these would be the people more likely to store things neatly. Then, some people may adopt a minimalist approach and seek to possess few things. Their closets would not only be orderly but also relatively bare.

Because memory is one's life experience recorded in the mind, how memories are stored tends to reflect one's lifestyle. If you live a chaotic life, your memories will be stored haphazardly and may be difficult to access. If you live a standardized life, your memories will be stored in an orderly fashion and will be more easily accessible. And if you simplify your life completely, your memory will not only be orderly but also relatively uncrowded, meaning you will have an exceptionally clear memory. When we get into the function of the memory in just a moment, we will return to this concept of accessibility.

Earlier, we discussed the concept of reincarnation—the process by which the subtle body transitions from one gross body to another after death and before birth. Because memory is a component of the subtle body, it participates in this transition. Then one may wonder, why can't they access their memories from past lives? Aren't they contained in the memory? This is a valid question and we will explore it now.

First, rather than probing into your past lives, let's explore more a recent memory—as recent as the day before. How much of that experience do you remember? What percentage of your experiences and sensations do you recall? These memories *are* recorded; we know this is true because we often suddenly remember something we thought we had forgotten. But it is still difficult to access memories we formed less than 24 hours ago.

Next, consider the sheer number of past life memories that would be contained in your current memory. When we get to causal body analysis, we will discuss how we all possess infinite past lives. But forget infinity, imagine you had one billion past lives, or even one thousand, or one hundred. Can you imagine trying to remember that many lifetimes of experience? That many spouses, that many children—every single person and thing and sensation and experience that all those lives contained?

The memory has systems that help us forget—the equivalent of purges and defragmentation. Just as in this life, you tend to remember only the more significant events, when we die and are reborn, we tend to retain access only to the most significant past life events as well. This process attenuates the memories over time; they are all still stored and still contribute to the cumulative resultant force that shapes us at birth, but our ability to access them diminishes. **Because memory contributes to the resultant forces that shape our subtle and gross bodies in each new life, it is desirable to possess a good memory so that we can have better insight into the factors that determine the nature of our existence.** This is why it's valuable to have an orderly lifestyle—to improve our access to our memories. Better yet is a minimalist lifestyle, which makes accessing memories that much easier.

As we conclude this discussion of memory, recall some of the indications that we do maintain some access to the memories of past lives. We noted how newborn babies sometimes smile in their sleep. This is because dreams are iterations of life experiences, which the baby is too young to have formed in this life. This indicates that newborns retain some part of their memory from their past lives.

We also mentioned child prodigies—very young children who display preternatural skills in music, mathematics, chess, spirituality, or any other field that they are too young to have learned in this life. Such prodigious skill is likely to result from past life memories that express themselves strongly enough to carry over into this life and inform the child's experience. Such indications contribute to our understanding of the function of memory, one of the main components of the inner personality.

The ego

The final component of the inner personality and the entire subtle body is the ego. **The ego is nothing more than a thought that recognizes that the subtle body and gross body that you currently possess is *you*. The function of the ego is to provide a constant point of reference as an individual goes through life. It establishes one as a subject, and everything outside of that as an object.**

While the term *ego* has a somewhat negative connotation in its common usage (which we will get to in a moment), in its fundamental form the ego is benign and inescapable. It simply exists to confirm that a given gross body-subtle body combination *is* that given individual. It's a simple, factual declaration—nothing more. When the benign form of ego gets more pronounced, however, it can cause damage to the personality.

The ego can start becoming too pronounced when it begins to overlap with memory. Consider what it means when we say someone is egotistical or that they have a big ego. It generally means that they put too much stock in the things they have achieved, and have begun to identify their *self* with their past experiences. Their declaration of self has become linked with their memory.

Now, our experiences are factual. They happened, and the fact that they happened can't be erased. They exist in the memory, regardless of whether we can recollect them or not. The ego is also factual, in that its existence is inescapable. Those two premises should remain separate. When a person begins to conflate them, that is when the ego becomes excessively pronounced, beyond the scope of its normal function. Note that this does not only take the form of bragging about how great one is, or the things one has achieved—the common understanding of the term *egotistic*. Inferior egos—possessed by those with inferiority complexes—can also be damaging, both to oneself and others. The ego is dangerous whenever it is excessively pronounced; the direction of the deviation doesn't matter.

This concludes our discussion of the 20 components of the subtle body, along with their respective functions. As a whole, the function of these 20 components is to direct the actions of the gross body in the physical world. They allow us to respond to stimuli, conduct transactions, and experience all the pleasure and pain the world has to offer.

We will conclude our discussion of the subtle body in the next chapter, which will deal with its nature. Then, we will move on to explore the causal body—a concept we have only mentioned but have not yet probed.

Chapter 9

Human Composition: Subtle Body - Part III

Having discussed the materials, components, and function of the subtle body, let us now complete our analysis by describing its nature. This nature roughly corresponds to the four main characteristics of the gross body's nature—but with some fundamental differences.

The nature of the subtle body

The first characteristic of the subtle body is that **like the gross body, the subtle body changes continuously.** However, its rate of change is much slower. Changes in the subtle body are generally directed by its controlling force—the intellect—which determines what changes should be made and how to carry them out. Sometimes, the mind can act on its own, leading to changes based on its whims and fancies. The intellect recognizes these changes when they occur, and depending on its strength, may or may not take action to override the mind's changes.

A reasonably strong intellect will be able to override the mind. A weaker intellect will notice the mind's changes but will remain an observer. For example, a person's mind may direct them to take actions that lead to unhealthy outcomes. The intellect knows these behaviors are unhealthy and knows which ones would be better. However, if the intellect is not sufficiently strong, it will not intervene, and the person will do nothing about their unhealthy lifestyle. A more powerful intellect will be able to overcome the mind and, in this example, may direct the body to start exercising, or follow a better diet. In other words, it will direct new changes to override the changes of the mind.

The changes of the subtle body unfold over long periods of time. Extending our example, a person can't become fit overnight; their gross body takes time to change and so the subtle body must direct that change over a longer period. Even after the gross body has completed its change, the subtle body must continue to maintain the gross body's new state. Therefore, the changes of the subtle body are necessarily of longer duration than those of the gross body. So, one aspect of the subtle body's nature is that it is subject to change that takes place slowly.

The next aspect of the subtle body is its limited lifespan. Previously, we established that each gross body has a maximum lifespan of roughly 120 years. For most, this lifespan is much shorter. The subtle body remains united with the gross body for the duration of this lifespan, but after its death, moves on in search of a new gross body.

How can we then say that the subtle body has a lifespan? The answer is that the lifespan of the subtle body stretches from the Big Bang up to the Big Crunch or the Big Rip—the total dissolution of the universe. These topics will be discussed in detail when we eventually get to the topic of the Big Bang and creation.

The subtle body first manifested at the time of the Big Bang, and its cycles of rebirth will continue till the universe is dissolved. As the universe itself is 13.8 billion years old, that is the current age of the subtle body. How much longer will it last?

For many more billions of years, potentially however long it takes the universe to reach the point of total dissolution. At that point, the subtle body will dissolve as well.

You may wonder how the subtle body can be so old when human beings didn't exist until very recently. The oldest fossil belonging to the genus *Homo* dates back to roughly 2.75 million years ago, and the oldest *Homo sapiens* fossil is only about 300,000 years old. But this takes only the fossil record on our planet into account. Life forms inhabited by the subtle body may have been present earlier elsewhere in the observable universe.

Regardless, the Big Bang occurred 13.8 billion years ago, and gross bodies capable of uniting with subtle bodies did not appear until later. So what were the subtle bodies doing until the gross bodies appeared? When we get to the chapter on creation, we will explore the concept of a subtle universe emerging from a subtle Big Bang, and discuss how the subtle body behaves in relation to those concepts. But that is a topic for later.

In brief, though, it took time for the optimal circumstances to arise for the subtle body and gross body to unite. Try drawing a comparison to cooking rice. When you first put the rice and water together in the pot, nothing happens. It is only after the right level of heat has been applied for the right amount of time that the rice cooks properly. Similarly, it took the universe many billions of years to produce the optimal circumstances for uniting the subtle body with the gross body in human beings. But the subtle body nonetheless existed long before that. It will continue to exist, traveling from one gross body to another, until the total dissolution of the universe.

The final aspect of the subtle body has to do with perceptibility. While the nature of the gross body is perceptibility, the nature of the subtle body is imperceptibility.

The gross body is perceptible to both ourselves and others. The subtle body, on the other hand, is not perceptible to others. It is only perceptible to ourselves. We register and experience our thoughts, sensations, and emotions—but the thoughts, sensations, and emotions of others remain inaccessible to us. It is as if the subtle body is protected by a highly encrypted password, and there's only one person capable of unlocking it.

Neglect of the subtle body

This completes our analysis of the subtle body, encompassing its raw materials, components, function, and nature. And yet, despite the accessibility of this information, so many people remain ignorant of its complexity. Do you recollect the image of the person standing before that enormous pumpkin and denying its existence? This is the state of affairs regarding the subtle body for most people, even the rare 10 million engaged in the pursuit of infinity.

At most, people will acknowledge the mind and may entertain some notions of the intellect, memory, and ego floating around inside it. But they will refuse to understand, or even acknowledge, the existence of the 20 components of the subtle body, let alone their function and nature. As we discussed in the second chapter of this book, too many people conflate terms that, in reality, refer to distinct concepts and do not recognize their actual definitions and functions. These essential components of the subtle body are not studied or taught in psychology and psychiatry either.

While these subtle components are bound with ideas of reincarnation and pure states of matter, it's not necessary to believe in those things to acknowledge the configuration of the subtle body. Of course, we *will* substantiate those ideas later in this book; but for the time being, the fundamental definition of the subtle body doesn't require such an understanding.

The unfortunate thing is that the subtle body is not even all that difficult to understand. Complex topics pertaining to the gross body are already taught in schools and universities all over the world. Scientists, researchers, and medical professionals are making astonishing accomplishments as they probe deeper and deeper into the gross body. Why then, should the subtle body be completely neglected, when the other is so thoroughly embraced and explored?

It's not that people aren't capable of learning. High school students study complex trigonometry, algebra, calculus—even the basics of quantum physics. If they can understand these subjects, it should also be easy for them to understand the components, function, and nature of the subtle body. But most people are not taught these subjects. Even if 10 percent out of the rare 10 million intellects had an awareness of these topics, it would be a high number.

Therefore, as you work through this book and continue to enrich your understanding of the subtle body, recognize that you are acquiring an insight that is extremely powerful and rare. As such, consider it your responsibility to pass this information on to others—particularly the younger generation who will have more time to learn and apply this insight in their own lives and work.

People love doing things that allow them to grow their intelligence and increase their capacity. They read self-help books and attend lectures, seminars, and conferences geared toward personal and professional growth. People study stress management, time management, relationship management, and so on.

In reality, all these things boil down to the intellect managing the mind, and the mind managing the body. So a clear understanding of the relationship between the subtle body and the gross body—and, specifically, the relationship between the intellect, mind, and gross body—would be an invaluable asset to any form of human occupation.

As we discussed in the previous chapter, access to information—to good intelligence—is only half of the equation. A strong intellect is also required. How does one strengthen one's intellect? The same way one would strengthen a muscle—by nourishing it and by pushing the limits of its existing strength.

One nourishes one's mind by taking in a steady diet of good intelligence—but one must push the intellect to its limits to increase its strength. The intellect can also be compared to a knife. If a knife isn't used for a long time, it becomes rusty and dull and will no longer cut well. Similarly, the intellect must be used regularly to prevent the mind from taking over. At the same time, as knives naturally become dull with use, the intellect also becomes dull when it uncritically repeats the same actions over and over. It can fall into a rut. Knives must be sharpened periodically to restore their edge. Similarly, the intellect must be sharpened by testing it against new thoughts and new activities. This will not only keep it in good working order but also increase its strength over time.

This brings the analysis of the subtle body to an end, and positions us to advance our study of the human composition to the third and final body—the causal body—to which we will devote our attention over the next two chapters.

Chapter 10

Human Composition: Causal Body - Part I

Foundations and Four-Factor Analysis

In the last chapter, we provided some foundations in subtlety before beginning our analysis of the subtle body. In this chapter, we'll take the same approach. Before we get into our four-factor analysis of the causal body, let's establish some foundational concepts. This subject will not be treated in detail until later chapters that cover it explicitly, but some background will be valuable in understanding this complex concept that has not yet been well defined.

Foundations in causality

Let's begin this discussion by posing a few questions about the existence of gross and subtle matter at various points relative to the Big Bang. You will recall from the previous chapters that the gross body has a fixed lifespan that extends up to a maximum of roughly 120 years and that the subtle body has a fixed lifespan that lasts for the duration of the universe.

That being the case, we are left to wonder: Where did all that gross and subtle matter come from? All the gross matter that makes up the entirety of the observable and unobservable universe, and all the subtle matter that makes up the subtle bodies of all beings—where was it prior to the Big Bang, and what will happen to them after it?

Let's break this down into two specific questions. First, what was the state of all gross and subtle matter, including the gross and subtle bodies, prior to the Big Bang?

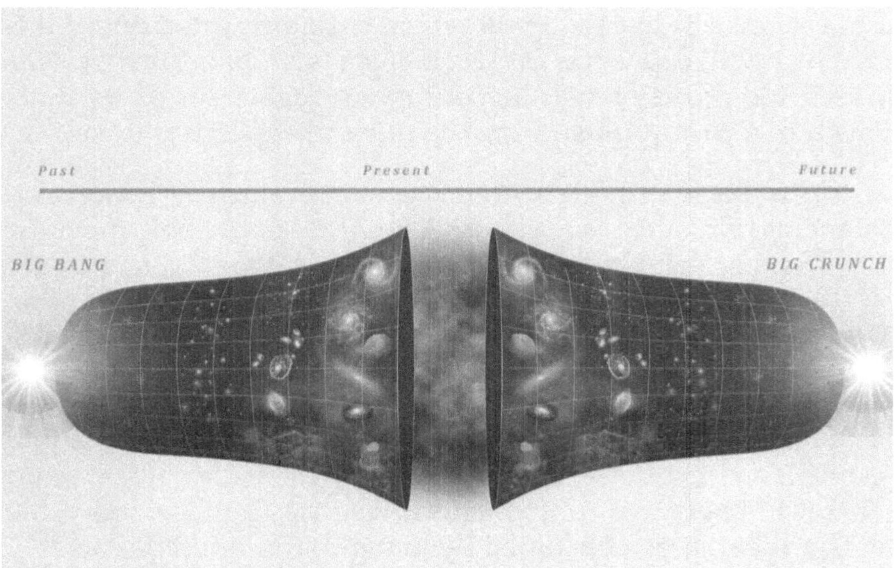

And second, what will be the state of all gross and subtle matter, including all gross and subtle bodies, following the total dissolution of the universe—the Big Crunch or the Big Rip?

By answering these questions and conducting an analysis of our answers, we can better understand causality and the causal body.

Conservation and the universe

To answer these questions, we can return to a basic, high school-level lesson—the laws of conservation of matter and energy. These laws hold that matter and energy can neither be created nor be destroyed, only transformed. Let's briefly run through some examples, which may seem simple but will help establish a foundation for our understanding.

First, consider a sealed system containing 100 cc of liquid water. If the liquid is boiled, it will evaporate and return to steam; if the steam condenses, it will return to liquid water; and if it is frozen, it will turn to ice. In all cases, the total quantum of atoms in this 100 cc water will be conserved, regardless of its state. This is one way of illustrating the principle of conservation.

Think back to high school chemistry, and the lessons on balancing chemical equations. Chemical equations must be balanced because, in all chemical reactions, the total amount of all substances involved remains the same both before and after. This is another example of conservation.

Likewise with the food all living beings consume. Regardless of the form, all life takes in some form of fuel, converts it into usable and unusable components, excretes the waste, and utilizes the nutritive components as energy. In this manner, the matter and energy consumed by living things are conserved.

To think of this in a slightly different way, consider a jeweler creating some golden ornament. Despite creating a completely new object, he doesn't create the gold itself. He has simply transformed the material into a new shape while conserving all of the original substance. At the same time, its original form as a gold ingot has been destroyed.

The same could be said of plants and trees. Even the tallest redwoods start off as tiny seeds that contain the potential to become a tree. Once the seed sprouts and becomes a tree, the seed is destroyed, but the tree is created.

Let's return now to the gross and subtle matter of the universe. These things cannot come from nothing. That would defy the laws of conservation of matter and energy. The Big Bang could not have 'banged' from nothing. *Something* that preceded it contained the matter that comprises our universe or the potential to become that matter. Therefore, we must conclude that, prior to the Big Bang, the entire universe existed in some unmanifested form. This unmanifested or potential form is what people refer to when they invoke the terms *singularity* or *potential universe.*

By definition, this singularity would have included everything that now exists in the present universe. All the gross and subtle matter would have been present in an unmanifested, potential form—a seed form, essentially. This is a state of no information or rather a state in which all information is present but highly encoded. In any event, these are the conditions that immediately preceded the Big Bang.

In that respect, the term *creation* (which we have used multiple times now and will be dealt with in detail later) is something of a misnomer—because nothing can be created and nothing will be destroyed. The entire universe will only and always exist in either a potential form or in the manifest form it occupies now. This cycle will continue eternally. Once our present universe reaches its moment of dissolution, it will resume its unmanifested form, only to manifest later at some point.

We can imagine an equation with 'creation,' manifestation, and expansion on one side, and dissolution, lack of manifestation, and contraction on the other. Universality and singularity. Nothing is created or destroyed as the state of the universe cycles between one and the other; the universe and its contents only transform from potential to manifest forms and back again, continuously and in perpetuity.

Conservation and human composition

As the purpose of this chapter is to introduce the concept of the causal body as part of the human composition, let's zoom in from the scale of the universe to the scale of the individual body. Based on the analysis above, we can infer that prior to the Big Bang, our own gross and subtle bodies must have already existed in the seed form.

That seed form represents the cause for your current gross and subtle bodies. **Therefore, the seed form of your total individuality can be referred to as the *causal body*. The matter of your gross and subtle bodies in their potential forms can be referred to as *causal matter*.**

The causal body is an entity that carries the gross and subtle bodies in the potential form during the period between universal dissolution (the Big Crunch or Big Rip) and universal creation (the Big Bang).

Before we continue, please note that the topics of causality and creation will be covered extensively in the coming chapters. But the present discussion must include at least some information on these complex topics. We are glossing over some major topics intentionally to prevent getting overwhelmed. Every subject will be treated in full in due time.

Now, let's continue with the causal body. We previously established that the subtle body is subtler than the gross body because it both controls and pervades it. Because the causal body contains the potential for manifesting both the gross and subtle bodies, it can be said that it both controls and pervades those bodies—and thus, the causal body is subtler than the subtle body.

This causal body, and the seed state of existence that it represents, have thus far escaped the study of the scientific community. Researchers have been focused on analyzing the creation of the universe—specifically the gross matter of the universe, both the observable and the unobservable forms. Their study has led them to the discovery of the quantum realm, containing quarks, leptons, and fermions—the new standard model of particle physics.

However, their research has not gone beyond the study of particles. There has never been a significant attempt to untangle the origin and nature of the universe in the context of subtlety and causality. Explanations of the universe generally depend on an objective, external study and analysis. Many refuse to even entertain the notion that there could be another way to approach these mysteries.

The same is true of the study of the body and mind. Such a study is generally confined to anatomical analysis. Even explorations of the mind and intellect conducted by neuroscientists and psychologists focus on consciousness as no more than a phenomenon generated by the gross organ of the brain.

Significant progress has been made in the scientific study of gross matter and the gross body, but the study of subtle matter and the subtle body is in shambles. With essentially no understanding of these things, the scientific community at large is not in a position to grasp the concepts of causal matter and the causal body, let alone study it productively.

With this introduction, we have established a solid foundation to advance our analysis of the causal body—the last phase of our analysis of human composition and a key component of our understanding of infinity and how to capture it.

Four-factor analysis of the causal body

With the above foundation in place, conducting our four-factor analysis of the causal body will be much more straightforward. In fact, by building on the lessons we've learned in the course of our study of the gross and subtle bodies, this analysis will be quick and intuitive.

We begin, as always, with the raw materials. We have established that the gross body is made of gross matter and that the subtle body is made of subtle matter—combinations of the five fundamental states of pure matter. Because the potential for the gross body and the subtle body are contained within the causal body, we can also conclude that **the potential for gross matter and subtle matter are also contained within the causal body. Here, they exist as causal matter—that is, as gross and subtle matter in their potential seed form.**

Causal matter is the subtlest form of matter. Beyond it lies the realm of non-matter or consciousness. The upcoming chapters on consciousness feature a beautiful derivation that illuminates the relationship between matter and consciousness—but that is a subject we will arrive at later. For now, simply understand that causal matter is the raw material of the causal body.

Next, what are the causal body's components? There are only two of them—the gross body and the subtle body in their seed forms. Just as a seed contains the potential for a tree, and thus is composed of that potential form, the causal body contains the potential for the gross and subtle bodies as its components.

Let us now define the function of the causal body. **The causal body's primary function is to serve as a storehouse for the gross and subtle body in the period between universal**

dissolution and manifestation. At the time of the Big Crunch or the Big Rip, the causal body receives the gross and subtle bodies, and it releases them again at the time of the Big Bang. A corollary of this function is that the causal body serves to sustain the interconvertibility of matter and energy, as it enables the infinite cycle of manifestation and unmanifestation of the gross and subtle bodies.

Finally, we'll wrap up this discussion with the nature of the causal body. Its nature is split into the same segments as those of the gross and subtle bodies. We will touch on each of them in turn, starting with perceptibility.

Recall that the gross body is perceptible to both oneself and to others, while the subtle body is perceptible to oneself alone. **The nature of the causal body is that it is not perceptible at all—not even to oneself.** We will return to this aspect of the causal body at the beginning of the next chapter because it has a special significance in the context of the next phase of our broader discussion.

The next aspect of the nature of the causal body is its lifespan. **The lifespan of the causal body is beginningless—it is essentially infinite.** What does this mean? Take a look at the image, which represents the cycle of manifestation and unmanifestation of the universe. One Big Bang is followed by a period of expansion, attenuation, dissolution, and unmanifestation, leading up to the next Big Bang. The same cycle is represented in the image as a beginningless, essentially infinite cycle. Note that we use the term *beginningless,* rather than *endless.* This begs the question, can this continuous cycle of birth and death end? And, if so, when?

In short, the cycle ends when one captures infinity. The causal body is extinguished at this time, and the cycle of birth and death comes to an end. This concept will be articulated at greater length in later chapters.

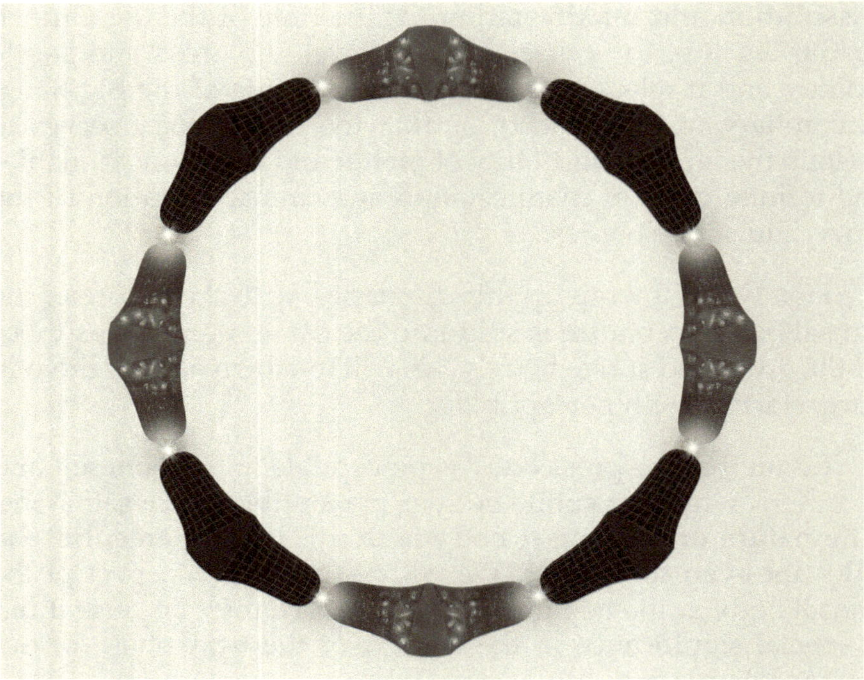

This point revives our conversation regarding the logic of capturing infinity. Recall from previous conversations that suffering is inherent in life because the first three categories of human goals contain inherent defects that cause pain and suffering even when one succeeds in attaining them. This same experience is what awaits each of us in every single lifespan, and every new manifestation of every new universe. This encompasses trillions of life cycles, stretching both into the past and the future, that are marked by constant pain and suffering. **This cycle of change and rebirth is another aspect of the nature of the causal body.**

Capturing infinity is the only way to escape this cycle.

Chapter 11

Human Composition: Causal Body - Part II – Further Analysis

In this chapter, we will continue our analysis of the causal body a little further, revealing a few additional aspects of its nature. These will have special significance as we gradually shift our discussion back to the pursuit and capture of infinity.

To begin with, recall one aspect of the causal body we mentioned in the previous chapter—its general imperceptibility. Unlike the gross and subtle bodies, which are perceptible to some extent, the causal body is wholly imperceptible, even to oneself.

We compared the causal body to a seed that eventually realizes its potential and becomes a tree. Now, imagine the tree wondering where its seed is. Of course, the seed is not present now—it has released its potential and become the tree. The seed can no longer be identified because its phase in the life cycle of the tree is over. In other words, *the tree is the seed* in its current form.

In the same way, it is not possible for a human being to locate the causal body. Why? To even pose this question, one must use one's intellect—which means that one must possess a subtle body and a gross body. If you possess them, it means that your causal body has already discharged its potential and no longer exists anywhere. *The current gross and subtle body is the manifested causal body.*

This inherent lack of perceptibility is an important aspect of the nature of the causal body, and it will lead us to several additional aspects that all bear similar importance.

Ignorance, infinity, and the causal body

As we spend the remainder of this chapter exploring various aspects of the causal body, we'll focus on the subject of ignorance, specifically as it relates to knowledge of infinity.

To begin with, recall our discussion of the subject-object relationships from a previous chapter. Each individual is a subject within a universe of objects. The subject can perceive these objects, study them, and gain knowledge based on their experience. The gross body takes in sensory perceptions, and the various parts of the subtle body analyze and respond to them. It is only when a subject and object interact that knowledge can be gained. This state depends on one possessing both gross and subtle bodies.

The moment these bodies revert to their unmanifested conditions—that is, to the causal body—they enter a state of total ignorance. This is because there is no possibility of a subject-object relationship in that state. Gross and subtle bodies exist only as potential; they are not capable of exerting any of their normal functions, so they become totally ignorant.

This state of total ignorance necessarily includes a total ignorance of infinity. The flip side of this is **once one has gained knowledge of infinity, the possibility of total ignorance will be destroyed.** Now, what does this mean in the context of the causal body? And what does *that* mean in the context of capturing infinity?

Capturing infinity is the end of the causal body

In short, this means that **the end of ignorance of infinity is equal to the end of the causal body.** Until that happens, its lifespan is a continuous cycle of birth and death. Once it happens, however, one is liberated from this cycle and merges with infinity.

Let's back up a moment to reflect on what this means for the causal body. **Part of its nature is the state of total ignorance.** This implies that **the causal body is also indescribable and indefinable. It is a state of no definition and no information—otherwise known as *singularity*.**

Why is it impossible to define this state? It's because we cannot claim that the causal body exists or does not exist. It is a state of total ignorance, about which nothing can be said. An example will help illustrate this point.

The state of ignorance is often compared to darkness. A dark room may contain any number of objects, but because there is no source of light, these objects cannot be perceived. The moment light is introduced, the darkness vanishes into nothing. Darkness has effectively been destroyed.

Now, for the most part, any substance that is destroyed will leave some remnant. However, darkness vanishes into nothing. Thus, one cannot prove that darkness is a positive substance. This may lead us to describe darkness as a substance or entity that does not exist.

And yet darkness *does* exist. Standing in that dark room, you know that objects surround you and yet you cannot see them, because of the darkness. So, darkness must exist, because you can both experience it and because it prevents you from experiencing other objects.

The causal body is like darkness. It both exists and yet does not exist; thus, it cannot be said *whether* it exists or does not.

We have now compared the causal body to both ignorance and darkness; let's continue these analogies a little further. Consider a completely dark cave that has not been explored for millions of years. It has been completely dark for that entire time. However, the moment you introduce a small spark of light into it, the darkness disappears instantaneously. Likewise, **the moment one gains knowledge of infinity—that is, *captures infinity*—their ignorance is instantaneously destroyed, no matter how many trillions of years or cycles of birth and death they have experienced before that.**

The origin of ignorance

We have established that the cause of the gross body and the subtle body is the causal body. The next question is what is the cause of the causal body? However, we've also established that the causal body exists in the space between the unmanifested and the manifested, an ongoing cycle that has no beginning. So how then can something originate in a cycle? This is another complex issue that will be dealt with during the chapters on the Big Bang, creation and consciousness, but a simple analogy will help clarify this point slightly.

Consider the following scenario. It is twilight, and a father and son are walking home from the movie theater. The movie they saw featured snakes. Up ahead, there is a length of thick rope lying coiled by the side of the road. Because it's twilight, the boy can only perceive the outline of the shape, and he perceives it as a snake.

Now, it's not unreasonable for him to believe that he saw a snake. In low light, a snake and a rope would be very similar in appearance. But the mistake itself is not the purpose of this example. Instead, consider *why* he believed he saw a snake. *It was because the rope was there at all.* **If the rope had not been there, he would not have expressed his ignorance of what it was; thus it was only because the rope was there that he was able to be ignorant of its nature.** Therefore, we can conclude that it is in the nature of things to not only contain the ability to be understood as what they are but also the ability to *not* be understood as what they are.

Now that we have introduced the principle of ignorance, let's take a moment to connect it with an idea we have already discussed.

In a previous chapter, we touched on the concept of infinite consciousness, which lends consciousness to our subtle bodies.

For this infinite consciousness to truly be infinite, it must also be all-pervading. Also, based on the rope/snake example, if the infinite consciousness exists, then the possibility of not understanding also exists within it.

Three conclusions flow from these points. **The possibility of not understanding the infinite consciousness must be *as pervasive* as its existence; the possibility of not understanding infinity must be *as pervasive* as its existence; and, finally, there must also exist a state of ignorance that is *as pervasive* as infinity.**

Can the causal body be found?

Let's bring the conversation back to our analysis of the causal body. We know that a part of the causal body's nature is its imperceptibility. As deep, curious thinkers, however, it is natural to wonder if it is still possible to find some trace of the causal body in this world. But is there any way this can actually be done?

One theoretical possibility is that we could obtain a glimpse of it through further exploration of the origins of the universe. As resources like the JWST probe further and further into the history of our universe, it is tempting to imagine that its discoveries may reveal some trace of causality. But can we use our tools to peer *beyond* the Big Bang? To gaze into the singularity itself? This is not possible and we will not be able to glimpse the causal body in this way.

Another alternative would be to be present as a witness at the dissolution of the universe and observe the transition of the gross and subtle bodies back into the causal body. This is also not viable, because it will take billions more years for this dissolution to occur and the subtle and gross bodies will be dissolved at that time as well. No one will be left alive to observe and store the knowledge.

A third possibility some readers may entertain could be to rip a hole in or even entirely destroy the universe to hasten the moment of dissolution and learn these secrets right now. This is the type of thing that only happens in movies. Not only is it impossible, but for the same reasons as in the second scenario, we would not be able to learn anything from it as there won't be any observers left.

While these three possibilities may not be viable, there actually is one way we can glimpse what the state of the causal body may be like. It is not the thing itself, but rather the closest approximation of it in this manifest universe.

First, recall what the causal body is. It's the state in which the gross and subtle bodies have become utterly unmanifest and are totally imperceptible, even to oneself. **In this way, the experience of deep sleep mirrors the state of the causal body.** During deep, dreamless sleep, one loses all sense of the gross and subtle body. Deep sleep cannot be recognized while it is happening; one can realize that one was in deep sleep only after awakening and returning to one's gross and subtle bodies. Thus, the experience of deep sleep gives us the best understanding of the causal body. We will discuss the experience of deep sleep in more detail in the chapters on the three classes of experience.

The causal universe

As our analysis of the causal body comes to a close, let's take a moment to zoom out from the individual causal body to a broader appreciation of the causal state. This will be a useful lesson ahead of later chapters dealing with creation.

Let's first take a step back from the causal body and consider the gross body. Just as an individual gross body is made of gross matter, the entirety of the tangible universe also consists of gross matter.

Now, what about the subtle body? You, the reader, have a subtle body as does every other living being on this planet—not to mention all the living things that may exist on other planets. Beyond that, inert objects which lack a subtle body also possess a subtle power of inertia that enables them to exist. This power exists within all the inert matter throughout the universe. (We will discuss this provocative idea in further detail when we discuss creation.)

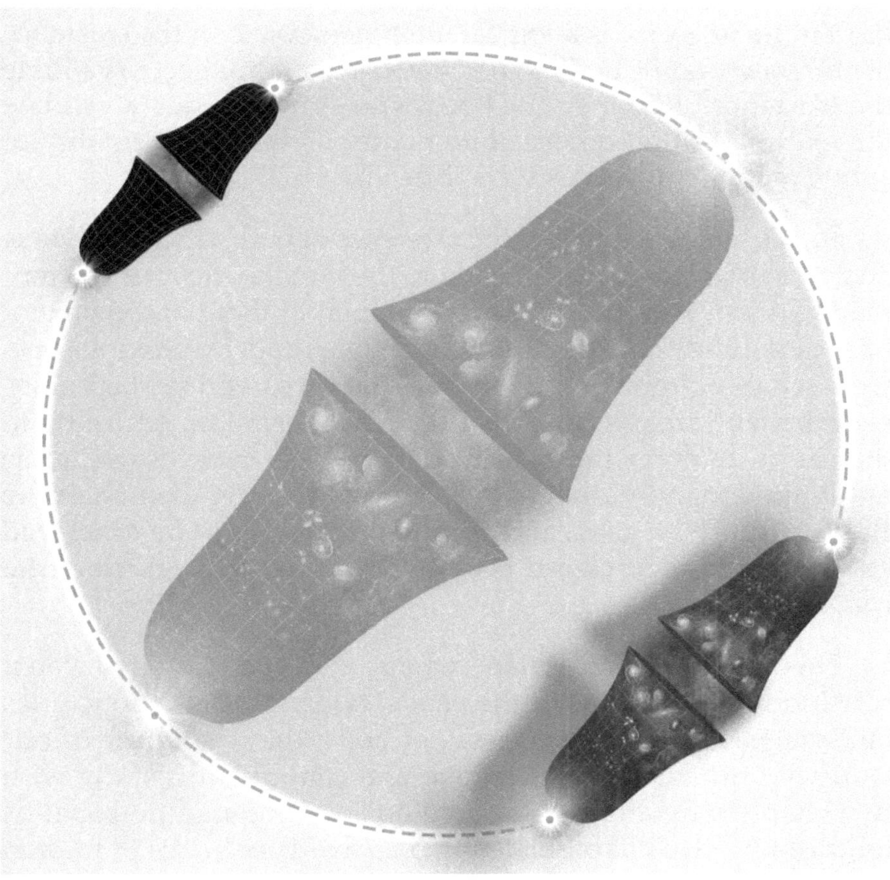

What does this mean on the universal scale? Gross matter is easier to understand in this respect; as it is perceptible to all. We can easily understand how all gross matter belongs to an interconnected universe of gross matter. But the subtle body is only perceptible to oneself and not others. If the subtle body is limited to one's own experience of it, can it also be a part of a broader field of subtle matter spread across the universe?

In fact, it can and it is. **The manifest universe contains an imperceptible subtle universe.** This subtle universe includes all the subtle matter—all the subtle bodies and subtle inertia of all beings and all the inert matter in the universe. The existence of this subtle universe is a logical inference based on the existence of one's own subtle body. When you use your intellect to validate the existence of your subtle body, you simultaneously validate the existence of all other subtle matter as well. This totality of subtle matter constitutes the subtle universe.

So, we have established the existence of both a gross universe and a subtle universe. Together, they make up our current manifest universe. But what will happen to them at the time of total dissolution? Throughout this chapter and the previous one, we have thoroughly explored how the gross and subtle bodies cycle between manifestation and unmanifestation, returning to the causal body in the interim period. This cycle, however, is not limited to individual bodies. **During the Big Crunch or the Big Rip, both the gross and subtle universes will be dissolved and return to their causal states—or, rather, will return to the *causal universe.***

This far-reaching claim, which will be revisited later, concludes our analysis of the causal body. This also concludes our analysis of the three classes of body, the first phase of our study of human composition. We will continue this study with two chapters devoted to the five layers of human personality, followed by two chapters devoted to the three states of human experiences.

Chapter 12

Human Composition: Personality Layers - Part I

The First Three Layers

The five layers of the personality of the human composition essentially correspond to the three bodies we have spent the last several chapters analyzing. Before moving on to an analysis of each layer, let's take a moment to establish a baseline understanding of the relationship between these layers and the three bodies using an example.

Consider the way governments are organized. A government is generally divided into different branches—usually executive, legislative, and judicial. For this example, these branches are the equivalent of the three bodies of the human composition; they represent distinct and broad realms of function and nature. Then, within each branch, there are many departments, bureaus, agencies, etc. These correspond to the components of each body, in that they are responsible for actually performing the functions and fulfilling the nature of each body.

Governing the activity of each of these components are specific sets of laws and guidelines that give them the power to perform their functions and fulfill their natures. These laws and guidelines are the rough equivalent of the layers of the personality. With that said, let us now begin our study of these five layers.

Introducing the five layers of the personality

We have indicated that the five layers of the personality correspond to the three bodies. Specifically, one layer corresponds to the gross body, three layers correspond to the subtle body, and the final layer corresponds to the causal body.

The first layer, which corresponds to the gross body, is known as *the food layer*. This is the body's external or physical personality, and it governs the body's ability to both grow and decay.

The next three layers correspond to the subtle body. The first of these is *the vital energy layer.* Also known as *the acting power of the personality,* it holds the **power for activity**. It is not the power *of* activity, which resides in the subtle body's components; rather, it is the power *for* activity. Remember the example of the government; laws don't act—agencies do. The laws contain the power for action, which is then carried out by the agencies. This dynamic is reflected in each of the layers.

Next is *the mental layer.* This layer is the emotional or psychological personality and represents the power for desiring objects. Once again, it is not the power *of* desire, but the power *for* desire—the capacity to generate thoughts and aspire to obtain some object. This mental layer can also be described as *the desiring power of the personality.*

The third layer pertaining to the subtle body is *the intellectual layer.* This can alternately be referred to as the thinking personality, rational personality, or judging personality. It is responsible for the power for knowing things. As with the previous two layers, this can also be described as *the knowing power of the personality.*

The final layer of the personality is *the bliss layer,* which corresponds to the causal body. This layer holds one specific power—suppressing or overcoming the other powers. This can be described as *the inertia power, suppression power,* or *unconscious power of the personality*. It is exhibited both during the period of unmanifest *and* deep sleep, when all other human powers are suspended.

The same order of subtlety that pertains to the three bodies also pertains to the five personality layers. We know that the causal body is the subtlest body, followed by the subtle body and the gross body. Similarly, the bliss layer is the subtlest personality layer, followed by the intellectual layer, the mental layer, the vital energy layer, and finally, the food layer.

With this framework in place, let's begin our study of each of the five personality layers.

The food layer

The food layer is nothing but the external, physical personality and it contains the gross body's power for growth and decay. To analyze the nature of this layer, let's explore the life cycle of the gross body.

The gross body is formed by the food it consumes, which originates from the earth and ultimately returns to the earth. So, the earth is the starting point. It is composed of nutritive matter, which produces vegetation, which in turn produces food and is consumed by humans. Once humans consume this food, their bodies break it down and use it for any number of purposes—including the important purpose of reproduction. The bodies of future fathers transform their food into sperm and the bodies of future mothers transform their food into ova. This specific transformation is the most significant transformation that gross matter is capable of because it leads directly to the creation of new life.

When the sperms and the ova unite, they begin to form a new human gross body. Now, how does this body continue to grow? The mother continues to take in food and transfers the nutrition to the growing baby. After birth, this process continues, first during breastfeeding and as the baby grows it begins to consume liquid and then solid food. And food sustains the body as it grows to maturity. Thus, food is responsible for the creation, growth, and maintenance of the gross body.

Eventually, the body ages and the food intake slows down. Then, inevitably, the gross body reaches the end of its natural lifespan and dies. After death, the body is buried and becomes food for the microbes, which return the gross matter that once comprised the body back to the earth. In this way, the food layer of the personality contains the power for the gross body to grow and decay.

The vital energy layer

The next layer is the vital energy layer, also known as the acting power of the personality. **This layer pertains to 10 components of the subtle body—the five power centers of action and the five power centers of physiological function.** Again, this layer does not refer to any of the gross organs associated with these components, but to the power behind the gross organs associated with action and physiological function.

Now, how does this power manifest? Let's start with the power centers of physiological function. When the body takes in its two critical resources—air and food (including water)—these power centers are responsible for breaking it down, circulating it throughout the body, and eliminating the resultant waste. The amount and quality of the food one consumes will vary greatly from person to person and the quality of the air may also vary. These power centers function in the same way for all people and the vital energy layer is responsible for supplying the power for these functions to occur.

Beyond the basic functions necessary to sustain life, this acting power of the personality also drives the actions that the gross body carries out. The five power centers of action (associated with the organs of speech, manipulation, locomotion, excretion, and reproduction) carry out their functions by consuming the energy that is generated by the power centers of physiological function. Because they are driven by the vital energy layer of the personality, the five power centers of action also depend on this layer.

Thus, **just as these 10 power centers are bridges between the gross body and the inner personality, this layer acts as a bridge between the food layer and the mental and intellectual layers.** This relationship will be described in more detail in the following subsection on the mental layer.

The mental layer

As indicated above, the mental layer refers to an individual's emotional or psychological personality. **It consists of eight components of the subtle body. These are the five power centers of perception—the ears, skin, eyes, tongue, and nose—along with the mind, the memory, and the ego.**

The mental layer and the vital energy layer have an interdependent relationship. The eight components associated with this layer depend on energy generated by the components of the vital energy layer to function. At the same time, because the mental layer contains the power for desire, and desire is a prerequisite of action, the mental layer also affects the vital energy layer. So, the relationship goes both ways. In the next chapter, we will discuss how the mental layer and intellectual layer interact, a relationship that also goes both ways.

The mind constantly generates desires and passes them on to the organs of action to carry them out. This is part of a cycle that is constantly in motion. The vital energy layer powers the mental layer to generate desires, which passes them back to the former to fulfill them.

For most people, the mind continuously generates desires. But how many of those desires can a person ever hope to fulfill? Most people have limited resources to fulfill their desires, so the gap between their total desires and their ability to fulfill them is constantly growing wider. This is the perennial problem of the human race. Over time, the average human being ends up as a bundle of insatiable, uncontrollable desires.

Desires can be equated with fire, with the mind as its fuel. Just as fire will continue to spread as long as there is fuel, so desires continue to grow as long as the mind generates them. And because the mind is an inexhaustible resource, this fire will never run out of fuel. This is why the richest person in the world will always want to be a little richer.

So much for the mind—how do the other components we mentioned function in the context of this mental layer? First, the five power centers of perception are constantly taking in sensory stimuli, which are passed on to the mind for consideration. This is a key driver of new desires, as this external data can inform the mind of new objects and possibilities. But this is also driven by memory and ego because the information stored in them can serve as the origin of new desires in the same way that external data can.

Let's take a moment to note that desire, in itself, is not a bad thing. All four categories of human goals, *including the goal of capturing infinity,* have their origin in desire. Desires are raw materials, just as food is a raw material for the body. They can be refined and given a higher purpose when they are transformed into goals. This is the case up to and including the goal of capturing infinity, the noblest goal a human can set.

Such refinement requires the intervention of the intellect, which is governed by the intellectual layer of the personality. We will discuss this in the next chapter.

Chapter 13

Human Composition: Personality Layers - Part II

The Final Two Layers, and Beyond

Let us continue our analysis of the layers of the human personality with a discussion of the final two—the intellectual layer and the bliss layer. After that, we will put all five layers into the broader context of capturing infinity.

The intellectual layer

The intellectual layer is the knowing power of the personality, and it comprises nine components. Like the mental layer, it contains the five power centers of perception, the memory, and the ego—but instead of the mind, it contains the two intellects. These are the gross intellect, which engages with the gross universe, and the subtle intellect, which is capable of contemplating transcendental concepts, such as infinity. How can it be that seven of these components exist in both the mental layer and the intellectual layer? We will touch on that in just a moment.

First, as the intellectual layer is responsible for the power of knowing, it follows that nothing can be known *without* the involvement of this layer. For a desire to even exist, some amount of knowledge is necessary. Even if the intellect is not strong enough to guide the mind, the knowledge that exists within the intellect informs the mind as to what can be desired. ***To put the three layers associated with the subtle body into perspective, the intellectual layer knows, the mental layer desires, and the vital energy layer acts to fulfill those desires.***

The five power centers of perception take in external stimuli and pass them through the mind to the intellect. Internally, the memory and ego also provide stimuli. It is the role of the intellect to register these stimuli, determine what to do about them, and direct the mind accordingly. The power of knowing, housed in the personality's intellectual layer, enables all this activity.

Now, what role does this power play? In short, **it helps us to limit the number of unfulfilled desires we entertain. This, in turn, increases our happiness—which can help bring us closer to capturing infinity.** Because it is the nature of the mind to constantly generate new desires, without the power of knowing we would quickly be overrun by unresolved desires. Fortunately, thanks to this power, we can examine our flow of thoughts and determine whether or not something actually should be a desire. We need not—and, indeed, *cannot*—desire everything that we are aware of, and the knowing power of the intellectual layer allows us to limit it.

For example, at the age of 50 or 60, it is not possible to become a soccer star, no matter how much one may desire this. The knowing power can prevent this from becoming a desire, which would remain unrealized. Similarly, for a person who doesn't have much money, buying a Ferrari would not be a realistic desire, and the knowing power would be able to prevent it from becoming one.

As to the question of how the same seven components can belong to both the mental and intellectual layers, you must recall that the raw material of the entire subtle body is thought. Some desires may spring into existence without the permission of the intellect, while others only become desires after having been validated by it. Thus, any internal or external stimuli can flow to either the mind or the intellect. In the first case, all the involved components belong to the mental layer. In the latter, they belong to the intellectual layer. This is how those particular components can belong to both the layers of the personality.

Note, however, that even when the mind generates a desire, this thought will always register within the intellect to some degree. This is because *desiring is not possible without knowing.* The intellect may be too weak to control the mind, but it has participated all the same. This is an important point to keep in mind when considering the interrelation of these two layers.

The bliss layer

The fifth and final layer is the bliss layer—the unconscious personality. **This layer pertains to the causal body and, during manifestation, to the state of deep sleep.** In deep sleep, the other layers are *unmanifested;* that is, they don't operate as they do during the other states of experience. Thus, this bliss layer is described as holding the power of inertia, or a suppressing power, over the others.

When we say 'other powers,' we are specifically talking about the knowing, desiring, and acting powers. There is also the power of growth and decay, but this is not a true power in the way that the others are. This is because it pertains to the gross body, and is hence a kind of gross power. The others pertain to the subtle body, which is entirely composed of the flow of thought. In deep sleep, the unique suppressing power of the bliss layer suspends the other powers and puts the personality into a state of total ignorance. There is no subject-object relationship, and so there is no possibility of gaining or utilizing knowledge.

During this period, all the remaining layers enter a state of relaxation. The physical body gets a rest, the mind is rejuvenated, and the intellect is given a break. The entire personality is revitalized, and afterward, can perform its normal, waking functions—knowing, desiring, and acting—better. So, the cyclical activity of the bliss layer plays an important role in the proper functioning of the human body. Additionally, during this time all worry and anxiety are temporarily suspended. The state of deep sleep is a very, very happy state. This is one of the reasons why it is known as the bliss layer.

Note that the other powers are suppressed not extinguished or obliterated. During the time that the suppressing power asserts itself, the physical body is still engaged in a bare minimum of physiological activity. It must be, otherwise we would die in a state of deep sleep. Likewise, the mind suspends all activity, which is resumed upon waking. It must be so, otherwise we would lose all our desires every time we entered deep sleep. The same is true for the intellect. It sustains itself throughout deep sleep, otherwise when we wake at 8 a.m., we wouldn't recognize that we are the same as the people who went to sleep the night before at 10 p.m. Thus, we must infer that a thin, rudimentary fiber of the other powers is sustained during this period of suppression, especially a thin fiber of intellect.

However, though these rudimentary powers endure during suppression, we cannot claim that the subtle body is subtler than the causal body. The criteria of the hierarchy of subtlety states that for one thing to be subtler than another, it must both control and pervade it. In this case, the other powers neither control the power of suppression (because they are being suppressed by it), nor do they pervade it (because they only exist as thin, rudimentary versions of themselves). So, we can understand how the causal body is still subtler than the subtle body, despite the thin infringement of the subtle body into it.

There is one more important aspect of the unconscious personality. Consider what happens to the electricity in your house when there's a power surge. The circuit breaker trips to protect the wiring and appliances. To put it more plainly, under circumstances that are more extreme than the system can handle, the system shuts down to protect itself.

The suppressing power performs a similar function. For example, when a person has a serious accident, the suppressing power asserts itself and prevents the body from experiencing the full measure of pain. The same is true when someone is faced with serious emotional trauma that the mind, intellect, and ego are not able to handle. This can happen at any time but is particularly common at a younger age when the intellect is less developed. In these situations, the suppressing power will assert itself and cause the experience to bypass the mind and intellect. Because everything that happens in one's life is stored, these experiences generally wind up deep in the memory, or may even pass into the unconscious personality itself.

We can see evidence of this when people exhibit behaviors that cannot be explained by their immediate context—for example, when someone reacts angrily to some minor problem, or when someone breaks down and cries for no apparent reason. Since the suppressing power is not always strong enough to hold back deep memories or unconscious experiences, they may come out at unpredictable times. But the fact that this happens at all is evidence that the suppressing power has asserted itself at some point in the past.

With that, we conclude our analysis of the bliss layer of the human personality. Let us take a moment to contextualize what we have discussed over the last two chapters, and also look ahead to what awaits us in the ones to come.

The personality in the context of infinity

Over the next two chapters, we will analyze the final three aspects of the human composition—the three states of experience. As we have mentioned, these are the waking state, the dream state, and the deep sleep state.

After analyzing the three states of experience, we will move on to a discussion of consciousness, infinity, and the relationship between them. After that, we will continue to build lesson upon lesson until we establish the actual methods for capturing infinity.

At this point, the reader may be wondering why we have spent so much time discussing human composition—all these power centers, personality layers, and raw materials—when the goal of this book is how to capture infinity, creation, and the mysteries before the Big Bang. There are several reasons why this lengthy analysis is a critical part of the process.

For one thing, it is important to study these 11 aspects of human composition to definitively state that they are finite—that is, that they are *not* infinity. Recognizing where *not* to look for infinity will prevent you from making wrong choices in the pursuit of capturing it. Capturing infinity is a noble goal that takes a good deal of preparation. It doesn't happen overnight. Thus, there will be many opportunities to go wrong. Taking the time to establish the strongest foundation possible will eliminate many of them before they even occur.

Another reason is that to recognize your path, you have to understand your starting point. This requires an accurate assessment of your personality. Personalities are generally dominated by one of the personality layers described in the previous chapters. For most people reading this book, the intellectual layer is likely dominant. This will be an asset in your attempt to capture infinity.

Yet, while one layer may dominate, the other layers are present as well. Together, they all form one's total personality. Even the most intellectually inclined individual has significant emotional, active, and physical aspects. It is important to recognize all aspects of your personality, because if you ever lose your grip on your intellect, the other layers will likely assert themselves. This will not only affect the nature of your personality but will also affect your progress toward capturing infinity.

Therefore, not only will an analysis of your personality help you understand your path, but will also help you assert control over your personality's non-intellectual layers. We have established that the mind's tendency is to keep generating new desires, and to capture infinity, one must eliminate all unresolved desires *except for* the desire to capture infinity. Without a sound understanding of your personality, you will struggle to balance its layers and will struggle to eliminate extraneous desires. *With* a proper understanding, however, you will gain clarity that will enable you to reduce the gap between your fulfilled and unfulfilled desires. This will gradually enable you to capture infinity.

A sound analysis of your total personality will allow you to push back against the forces within you that tend to suppress your growth. It will enable you to place the food layer, the vital energy layer, and the mental layer in their respective places, with the intellectual layer as their director. You will arrange them in their specific 'orbitals,' to invoke an image from an earlier chapter. As this happens, your intellect will become stronger—a necessary prerequisite for capturing infinity.

With a strong intellectual layer, you will be able to perceive the other layers as objects. To see your mind and body as objects will inherently distance you from them. In other words, it will help you view them objectively. Becoming objective in this regard has many benefits. For one thing, you will reduce your mind's tendency to generate desires because you will identify less strongly with the personality layers that would otherwise seek gratification. You will also be less anxious, which will enable you to worry less, manage relationships better, manage your time better, and think more clearly. Understanding your personality will also help you understand that of those around you, which will improve your decision-making skills, and your relations with others. All these things will help you achieve goals more quickly and to progress along your path toward capturing infinity.

Finally, recall that, when we first defined the experience of capturing infinity, we equated it with the experience of achieving infinite happiness. We also indicated that, while the three other categories of human goals contain inherent defects, the goal of capturing infinity does not. Beyond that, the more devoted you are to this goal, the happier you are. Thus, the path toward capturing infinity—which is to say, toward infinite happiness—will also be characterized by happiness. In fact, if you are not happy during this process, then you are not doing it the right way. **So, understanding your personality based on this careful analysis is a prerequisite to experiencing the happiness that comes with pursuing infinity.**

This is why it's so important to take the time to understand the 11 aspects of human composition before approaching the subject of capturing infinity. A strong understanding of these aspects will help you develop a strong intellect, which is the key to everything that comes afterwards. With that said, let us move on to the final three aspects of human composition—the three states of experience.

Chapter 14

States of Experiences - Part I

The Waking State

The final three aspects of human composition are the three states of human experience—the waking state, the dream state, and the deep sleep state of consciousness. Together, these three states comprise all of human experience. Throughout your life, you will always be in any one of these three states.

Unlike the four-factor approach we used to analyze the rest of the human composition; here we will take a three-factor approach. **The first factor is the condition of the subtle body during each state**. Because all these states involve the subtle body, we must determine the condition of the mind, intellect, memory, and ego in each of them.

The second factor is the nature of the experience—that is, what is the actual experience of each of these states *like*? **The third factor is the predominant medium of each state**—which aspect of which body is most predominant during each state? We will move through this three-factor analysis as we study each state of experience, beginning with the waking state.

The waking state

We begin our analysis by considering the condition of the subtle body. **During the waking state, the subtle body is fully functional.** The mind and the intellect are engaged in taking in stimuli, registering them, disregarding them or turning them into desires and/or goals, and directing the physical body to act. The memory is also active, recording all the experiences. The ego is active as well, exerting its influence over the rest of the subtle body while also being influenced by it. The power centers of perception are engaged in capturing stimuli, and the power centers of action and physiological function are engaged in their normal activity as well.

The condition of the subtle body during the waking state is that it is fully functional and fully available to gather new experiences, generate new desires, obtain new knowledge, and perform new actions. The entire subtle body is engaged in the support of all this activity.

Now for the next factor—the nature of the experience. **The nature of this state of experience is external, physical, and objective.** External and physical because, in this state, one's subjective inner personality constantly experiences tangible, external things, and it does so using gross physical instruments. The gross matter of the universe is perceived by the gross sensory organs, translated into stimuli that are registered and considered as thoughts, and ultimately translated back into actions taken by the gross physical body.

This state is also objective because it is characterized by the recognition of objects, *including* one's own gross body. Because the subtle body is fully active, there is a strong subject-object relationship between the inner personality (specifically the intellect) and the objects around it, extending up to the observable universe.

The physical, objective nature of the waking state is largely driven by active coordination between the power centers of perception and their corresponding gross sensory organs. We can clarify this point by jumping ahead briefly and drawing a comparison between the waking state and the other states.

During the other two states, both of which occur during sleep, one (obviously) retains all of one's gross sensory organs. However, because the power centers of perception are temporarily suspended, the components do not interact. Thus, objective and physical experience is not possible. A sleeper may be bitten by a mosquito, or a loud noise may occur nearby, and yet they will not respond. A response only occurs if the stimulus is strong enough to reach the rudimentary personality layers that remain active during sleep. So, while the sensory organs are still available for receiving stimuli, the lack of coordination with the power centers of perception prevents the experience from occurring.

Can a similar disconnect occur in the waking state? They can indeed, though not in quite the same way. You may recall a time when you were deeply engrossed in some tasks and someone called your name, but you didn't notice. The sound waves still traveled through your gross physical ear, but they didn't register in your intellect. This was because the relevant power center of perception was effectively suspended while all your mental energy was focused on the task at hand.

However, this experience is not exactly the same as what happens during sleep. This is because the power centers of perception are always active during the waking state, even if they are temporarily disconnected from the sensory organs. To draw an extreme example, consider a person in a sensory deprivation tank. They are unable to take in any sensory stimuli, which means that their sensory organs and their power centers of perception are effectively disconnected. And yet, they are clearly still awake. Thus, it is the *activity* of the power centers of perception (and the rest of the subtle body) that defines the waking state, rather than the *act* of perception.

Of course, one doesn't live inside a sensory deprivation tank. One lives in the physical world. So, the waking state is effectively defined by the coordination between the power centers of perception and the gross sensory organs.

This coordination is what enables humans to perform complex physical actions. For example, when a doctor is performing surgery, their powers of perception are as critical as their powers of action. **This describes another aspect of the nature of the waking state—it is transactional.** Sensory stimuli translate into thoughts, which translate back to actions. This nature is not an aspect of either of the other two states of experience.

We will note one final aspect of the nature of the waking state before moving on. **During the waking state, one has full access to the three layers of personality associated with the subtle body—the knowing power, the desiring power, and the acting power. Awareness of these personality layers allows one to control them, based on the strength of one's intellect.** This is a unique characteristic of the waking state that is not available to the other two states.

Finally, let us touch on the third factor—the predominant medium. As we have just established, the nature of the waking state is predominantly physical and objective; thus, **the predominant medium of the waking state is the gross body.** The waking state is defined by physical transactions—sensory stimuli transformed into action. These require the activity of the gross physical body.

Now, there is a fine distinction to be made here. We indicated that the subtle body is fully available during the waking state, and said that it is specifically the full availability of the power centers of perception that sets the waking state apart from the other states. So why isn't the subtle body the predominant medium of the waking state?

The reason is that the coordination between the subtle body and the gross body enables physical transactions that are only possible during the waking state, and not during the other two states. Thus, while the waking state *does* require the activity of the full subtle body, it is *defined* by the activity of the gross body. This is why the gross body is considered its predominant medium.

The waking state, sensory experience, and the memory

Before moving on to the three-factor analysis of the next states, let's focus on one specific aspect of the waking state—the relationship between sensory experience and memory. This will be helpful during our analysis of the dream state in the next chapter.

We have established that the nature of the waking state is predominantly physical and objective and that it largely consists of the coordination between the sensory organs and the power centers of perception. Throughout the day, this coordination is responsible for translating sensory stimuli into the flow of thought to be utilized by the mind and the intellect.

However, the mind and the intellect are not the only components of the subtle body that interact with new experiences. The memory does so as well. The human memory is an extraordinary device, more powerful than any storage device in existence. It records every aspect of all of one's experiences—sensory stimuli, emotional responses, intellectual calculations and decisions, and even the involuntary actions of the physiological systems. This is its power. Note that it is even capable of storing records of things that have never happened, such as fantasies, imaginary scenarios, mental projections, and so on. In short, if it exists in the subtle body, it is recorded in the memory.

Drawing a comparison between the memory and storage devices may provoke the reader to wonder about the brain's role in the formation and storage of memories. After all, the brain seems to be the physical repository of memory. Remember that memory is a component of the subtle body, while the brain is a component of the gross body. As such, their functions cannot and should not be equated. The intellect and memory are expressed through the brain at the gross level.

Let's return now to the relationship between memory and experience. During the waking state, one is constantly *consuming* new external experiences. And, because the subtle body is fully functional and completely available, the inner personality is also constantly generating new internal experiences. The memory stores all of this.

The use of the word *consume* here is deliberate. Just as a person consumes food through the mouth (and thus experiences the sensation of taste), all the other sensory stimuli are also consumed through the relevant gross sensory organs. And again, the memory consumes all the internally generated experiences. This is happening constantly—all day, every day.

Let's continue to equate the consumption of experience to the consumption of food. What happens when a person attempts to eat a large amount of food? No matter how much they are capable of consuming, they will gradually slow down until they reach the point where they simply cannot eat any more. This is not just a psychological barrier; even if a person tries to continue eating at that point, their body would eventually burst and fail. So, there is a limit to the amount of food a person can consume. Perhaps one exception to this might be found in the example of the ancient Romans, who would eat to excess and then deliberately vomit to be able to consume the next course of delicacy. But this was to evade the natural limit of the gross body. The limit did exist for them, as it does for us today.

No such limitation, however, exists within the subtle body. There is no limit to the amount of external or internal stimuli the mind can consume. Every single experience penetrates through to the mind. The intellect has a hard time keeping up with so many thoughts. This is why the mind winds up generating so many new desires, the intellect does not have the chance to dismiss them.

We have already established that, with or without the intervention of the intellect, the mind directs the body to fulfill its desires. We have also established that the gross body has limited resources to make these attempts—in terms of money, time, health, and so forth. So, it is inevitable that many desires remain unfulfilled. They remain in the mind and keep increasing. And yet, each of these desires is clamoring to be fulfilled. As their number increases, the situation in the mind becomes increasingly chaotic, like a massive traffic jam. One or two cars may be able to get through at a time, but behind them, there are many, many more—far more than can possibly make it through.

What goes on in a person's mind under these circumstances? In a word, *STRESS*! Stress is nothing but the experience of acute mental agitation caused by the accumulation of unfulfilled desires, all of which are clamoring to be fulfilled simultaneously. Stress diminishes the intellect's ability to process desires efficiently. And so, the situation compounds and the mind remains violently agitated with unfulfilled desires. This is the result of the inevitable nature of the waking experience—constant consumption of external and internal stimuli.

Let's return to the comparison we drew between the consumption of food and the consumption of stimuli. If a person attempts to bypass their natural consumption limit, their body will stop digesting the food and will instead activate the power center of reversal, causing the body to vomit. Vomiting also occurs when the body takes in food that is tainted, undercooked, or that otherwise cannot be digested. This is a natural function of the human body.

An identical function exists with respect to the body's consumption of stimuli, which are transformed into thoughts in the mind. All these thoughts pass through the mind and enter the memory. Some thoughts are well-considered and complete; others may be incomplete, agitated, and disconnected from a fully formed experience. To clarify this, imagine you are attending an exhibition, but have only a limited amount of time to visit each booth. Thus, you can derive only a partial experience of each booth. This is what we mean by a lack of a fully-formed experience.

So, these incomplete fragments of thought enter the memory. Such thoughts are the equivalent of undigested food for the gross body. And, exactly like undigested food, the memory seeks a way to expel them to relieve its discomfort. **This expulsion occurs in the dream state. Dreams are nothing but the expulsion of incomplete fragments of undigested thoughts from the memory.**

After throwing them out, the memory can rearrange its storage more efficiently. This process is similar to computer defragmentation. However, note that, because the memory stores *all* experience, the thoughts it throws out are not eliminated; they are only temporarily externalized in the form of dreams. The thoughts will later be assimilated once more, either to be stored more efficiently or to be thrown out again in the dream state.

Now that we have established the relationship between waking experience, memory, and dreams, we are better prepared to begin our three-factor analysis of the dream state of experience.

Chapter 15

States of Experiences - Part II

The Dream and Deep Sleep States

We have completed our analysis of the waking state of experience and have described the relationship between experience and memory. Let's continue on this subject with analyses of the remaining two states. We'll begin with the dream state.

The dream state

Once again, the first stage of our three-factor analysis is to establish the condition of the subtle body during this state. During the dream state, the subtle body is only partially manifested. To be precise, **out of all the 20 components of the subtle body, only a figment of the memory is active during the dream state.** This is a singular experience in which *only* the dreamer participates, accessing memories stored by the waking state.

In the previous chapter, we indicated that dreams are nothing more than incomplete fragments of thought thrown out by the memory. However, because we are awake while making this statement, it is inherently biased by our perspective. What, then, is the experience of the dreamer while dreaming?

The dream state is effectively a waking state for the dreamer. Everything that occurs in it seems real, and the dreamer is capable of making judgments and taking actions just as someone awake. Now, how can dreams thrown out by the memory feel so real? Recall our discussion of the raw materials of the subtle body. All aspects of the subtle body are composed of flows of thought, which in turn are composed of subtle matter. Thus, all memory is composed of subtle matter and, as dreams are the product of memory, they are composed of subtle matter as well.

Now consider the nature of memories themselves—that is, what memories consist of. We have established that memories are full-spectrum, multi-dimensional recordings of everything a person experiences (both externally and internally), in this life and all the past ones.

The memory is not only a multi-dimensional recorder but also a multi-dimensional playback device. It follows, then, that its playback *also* presents a full-spectrum, multi-dimensional experience to the dreamer. This is why dreams include complex elements like space, time, emotions, thoughts, and so on. This is why we can say that, for the dreamer, the experience of dreaming is like a waking state.

Recall what we claimed at the beginning of this section—that, during the dream state, only the memory is active. How can such real-seeming experiences take place? **The answer is that the memory contains records of all other components of the gross and subtle bodies.** It contains a dream ego, a dream intellect, a dream mind, and a dream gross body. The memory knows how these entities experience things in the waking state and can replicate that experience in the memory itself during dreams. This is how the dream state can generate such real experiences, despite the memory alone being active during dreams.

Let's now examine the second factor—the nature of experience in the dream state. We saw that the nature of the waking state is external; conversely, **the nature of the dream state is totally internal.** As dreams only take place in the memory, all experiences are, by definition, internal. Externally, you may be sleeping in Boston, but internally you dream of being in Shanghai or Los Angeles. The external physical state has no bearing on the dream state, and the internal dream state has no bearing on the external. In the waking state, there are limits to what one can do and experience. In the dream state, the same limits do not apply. Multitudes of experiences can flow by in an instant and in a disjointed, jumbled way. This is because while the real world consists of gross matter, the memory consists of subtle matter—which is not bound by the same limits as the gross matter.

We also saw that the waking state belongs to the objective universe. **The dream state, on the other hand, exists in a subjective, private universe of the dreamer's own creation.** For the memory to manifest its dreams, it generates a dream universe in which they can take place. This is generated by the memory's multi-dimensional recording capacity—which is extremely powerful. However, its limitation is that experiences that occur there are confined within that subjective universe. In the waking state, one's actions can affect oneself and others; in short, this state of experience is transactional in nature. The dreamer may *experience* transactions, but they are dream versions only; they do not affect anyone or anything beyond the bounds of the dreamer's universe. And when the dreamer awakens, the dream universe is gone along with the dreamer. The same dream can never be revisited, perhaps one may get a similar dream again.

Let's examine another aspect of the subjectivity of dreams. Consider this statement: *It is not possible to dream what you have not experienced when awake.* Once again, dreams are only incomplete thought fragments thrown out by the memory. It follows, then, that the memory cannot throw out things that never went into it in the first place. At the same time, though, we have all dreamed about things that we have never actually done. So how do we account for this?

To explain this, let's draw on an example we introduced in the previous chapter—a somewhat unsavory example, but an effective one, nonetheless. Now, it's common knowledge that one cannot vomit up something that hasn't already been ingested. However, when one vomits, the matter ingested doesn't come out in an orderly fashion; everything has been transformed to some extent and is jumbled together unpredictably.

This logic explains why a dreamer will experience things that don't precisely correspond to actual events. Unfinished fragments of sensations, thoughts, experiences, places, and people all blend in unpredictable ways and can form seemingly new experiences in the dream state. But these jumbled experiences actually comprise real experiences from the waking life.

Even dreams that seem so novel or unusual that they couldn't *possibly* correspond to real events can be accounted for by this logic. This is because, when the memory throws out fragments of thought, it doesn't give precedence to 'real' experiences over imagined ones (even the imaginations and fantasies in waking states are also part of memory), recent ones over those from the distant past (or even past lives), or 'important' ones over those forgotten. The memories simply come out and combine, and dreams result from this combination. A long-forgotten memory, a scene from a fantasy like *Harry Potter* or the *Lord of the Rings*, and the knowledge of a single word may all combine to create something totally unique that has no direct precedent in the person's experience—yet it *is* based only on their past experience. Indeed, it must be, because it originated within the memory. **Thus, it is the nature of the dream state to create totally new experiences, based on memories from the waking state.**

To drive this point home, recall an example that we have referenced several times throughout this work—the example of a baby smiling during sleep. This expression of happiness can only be the result of a dream, which means that it is the result of a memory. However, because the baby has not had sufficient experiences in the current life to justify this expression, it is likely that the memory in the dream does not belong to this life but to the past one. As the baby grows and accumulates more experiences in this life, the total number of unfinished fragments of thought in their memory will grow as well, quickly outnumbering any past life memories. This explains both why babies can dream of their past lives and why it's uncommon (though not unheard of) for older people to recall their past lives in dreams.

Now for the third stage of our analysis of the dream state—identifying its predominant medium. This is precisely what we have been describing throughout this chapter. **The memory is the predominant medium of the dream state.**

Now, given that the gross body is the predominant medium of the waking state, and the memory is the predominant medium of the dream state, how does the transition from one to the other occur? We have noted that the dream state is essentially a waking state from the perspective of the dreamer, and the waking state is itself one. So how does one transition smoothly from one state to the other? The answer is that one has to essentially pass through the deep sleep state first, before transitioning into the dream state.

Throughout the day, the three powers associated with the subtle body drive the waking state of experience. These, of course, are the knowing, desiring, and acting powers. When these three powers exert their influence, one can't help but be in the waking state, because the mind is fully engaged and pushes the body to take action. However, the body must eventually rest. When one reaches this point, the power of inertia takes over and suppresses the other three powers. This induces sleep. Afterward, once the knowing, desiring, and acting powers have been suppressed, the dream state can commence. Later, the rudimentary versions of those powers that exist during sleep will pull the dreamer back to the waking state—though they must pass through the deep sleep state again in the process.

Thus, one must pass through the deep sleep state to enter the dream state. Likewise, from the dream state, one cannot immediately enter the waking state; one must pass through the deep sleep state again. This may seem hard to accept at first because particularly when waking up, it often seems as if one has passed directly from the dream state to the waking state. Recall, however, that both the dream and waking states involve perception and experience but the deep sleep state does not. Thus, in the interval between dreaming and waking, perception is suspended, meaning that one is totally unaware that this state has been traversed. This is why it may seem as though the change is direct and instantaneous though this is not the case in reality.

Now that we have nearly concluded our analysis of the dream state, let's explore in a little more detail a topic that we touched upon previously—the apparent reality of the dream state to the dreamer.

Earlier we noted that, for a dreamer, a dream is *not* a dream. From that perspective, the dream state is the waking state, and the dreams appear to be real. Of course, we understand now that dreams are only the memories of the person. But that fact is meaningless to the dreamer. To the dreamer, all the experiences are as real and valid as waking experiences are to the person who is awake.

Sitting in the waking state, we know that experiences in the dream state are unreal. The dreamer is ignorant of this and believes the experiences are real on their own. Recall now that the dream state is not bound by the same limitations as the waking state because it comprises subtle matter rather than gross matter. Thus, the dreamer believes that existence is in a waking state.

Therefore, to the dreamer, the unlimited potential that exists in the dream state is real. They may dream of distant galaxies; they may dream of a dream version of the JWST that is hunting for newer galaxies; they may dream that this telescope can reach the edge of the universe and capture the oldest galaxies it contains. All of that can happen in the dream state and it will be entirely real to the dreamer. It is only after waking up that they acknowledge the dream's unreality.

This completes our analysis of the dream state, an analysis that will be particularly valuable during our study of consciousness, infinity, and other abstract topics in the coming chapters.

The deep sleep state

The final state of human experience, and the final aspect of human composition, is the deep sleep state. We have already discussed this state in substantial detail during our study of the causal body. Let's apply the same three-factor analysis here as we bring our systematic study of the human composition to a close.

As to the first factor, what is the condition of the subtle body during this state? **In the deep sleep state, the subtle**

body is completely non-functional. We have discussed how the suppressing power takes over the three powers associated with the subtle body. These powers exist in the rudimentary, dormant states during deep sleep, and only resume activity once the waking state has resumed. It is only at this time that the person can register that deep sleep has occurred.

The second factor of our analysis pertains to the nature of deep sleep. **Even though the powers associated with the subtle body persist in a rudimentary fashion, the nature of experience in the dream state is total ignorance.**

We saw in the previous chapter that the nature of the waking state is external, concrete, and objective. In this chapter, we saw that the dream state is internal, abstract, and subjective. In the state of deep sleep, there is nothing more than total blackness and total ignorance. This total absence of stimulus and perception is what allows both the gross and subtle bodies to rest and recuperate. This state of total ignorance means that one is ignorant of their anxieties and worries. This is why the personality layer associated with this state is known as the bliss layer. The saying *'ignorance is bliss'* is an apt way of describing this state. Note that this state of bliss is shared by all people, no matter who they are. The moment they pass into deep sleep, everyone shares the same measure of bliss. From the people who spend thousands on luxurious first-class seats to those in economy, the moment they pass into deep sleep, it really does not matter how they travel.

It should also be noted that the state of deep sleep occurs during artificially induced states, such as a medically induced coma or when one is given an anesthetic for surgery. Despite being artificial states, the powers of the subtle body are suspended in just the same way that they are during natural deep sleep. The same criteria apply, meaning that the nature of these experiences is the same as the nature of true deep sleep.

Finally, what is the predominant medium of the state of deep sleep? We know that during deep sleep, the gross physical body is essentially in a state of suspension. Of course, certain involuntary actions do continue to take place—breathing, circulation, digestion, and so on. However, there is no association between these physical actions and the experience of the individual. That is to say, the gross physical body is effectively unavailable during the state of deep sleep. Therefore, it cannot be said that this body is the predominant medium.

We know that the subtle body is also almost entirely suppressed during deep sleep. Its powers do persist in an extremely rudimentary fashion—just enough to sustain continuity from the period before deep sleep to the period after it—but they are not capable of asserting themselves in any way. So, it cannot be said that the subtle body is the predominant medium of this state.

Thus, we can infer, by way of negation, that **the predominant medium of the state of deep sleep is the causal body**—the body that houses the gross and subtle bodies in potential form i.e. when they have not manifested yet.

With this, our analysis of the 11 aspects of the human composition is complete. As mentioned previously, this analysis will be critical during your attempt to pursue and capture infinity. It will also provide meaningful context for the remainder of our study here, which continues in the next chapters covering the subjects of consciousness, the Big Bang, creation, and infinity.

Chapter 16

You and Consciousness

Part I – Defining Terms

Having completed our analysis of the human composition, we are finally prepared to move on to a discussion of the nature of consciousness and its relationship with the individual.

Modern science and philosophy have produced compelling but inaccurate explanations about the nature and origin of consciousness. We will briefly touch on those explanations before continuing with the main thrust of our study.

Matter and consciousness

The concept of consciousness has been analyzed by scientists, psychologists, and philosophers alike, with compelling yet unsatisfactory results. Let's explore some of their theories and then contrast them with the true nature and origin of consciousness.

Modern theories of consciousness

We will begin with the scientific community. **Scientists, for the most part, attribute consciousness directly to the brain, claiming it is a physical phenomenon.** Those who hold this belief contend that consciousness is generated by the brain itself. This perspective focuses on the study of the brain's physical components—the cerebrum, cerebellum, medulla oblongata, neurons, axons, synapses, neural networks, and so on. The conversation also includes aspects of the nervous system—the somatic, autonomic, and enteric systems. Scientists ascribe the origin of the human consciousness to these physical components.

All these components are made up of gross matter. Thus, they are exactly the same in a living body as they are in a recently deceased body. That is to say, the atoms of the brain and nervous system in a living body are identical to those in a recently deceased dead body, which does not express consciousness.

How can it be, then, that consciousness is generated by, and is part of, the physical body? Science insists that consciousness is generated by the brain, and yet has never been able to validate this assertion. This lack of validation reveals that the claim is no more than a guess.

Of course, this is not to denigrate scientists for their current lack of evidence. All science is based on the exploration of the unknown and begins with incomplete information. There is still so much more to learn about the brain and, with the advent of artificial intelligence, we are entering a particularly interesting period for the study of the brain, thought, and consciousness. For now, however, we must acknowledge that the brain is a gross physical organ and that there is no evidence that consciousness originates in it.

Now, what theories have other thinkers proposed about the origin of consciousness? **Some psychologists and philosophers contend that consciousness originates in the mind, or even that it *is* the mind.** This, at least, gets closer to reality in that it doesn't claim that consciousness originates from gross matter. All the same, we know that this explanation is doomed to inaccuracy because it is not based on a correct understanding of the nature of the subtle body. Without an understanding of the subtle body, there is no way someone could develop a proper understanding of consciousness.

We know that the subtle body consists of 20 distinct components and that the mind is only one part of this group. And yet when people discuss the mind, they often do so as if the mind is its own distinct entity, with no broader context. *They may mention a body-mind-soul framework—another highly commercialized concept.* They may also talk about the memory, ego and intellect, but generally suggest that these components exist *within* the mind, rather than as distinct parts of the subtle body. With this faulty foundation, there is no basis for a proper understanding of what consciousness is, or where it comes from.

So much for what consciousness is not, how are we to understand what consciousness *is*? To achieve this understanding, we must first return to some of our previous lessons—starting with our analysis of the three bodies.

Matter borrows consciousness

We first studied the gross, physical body which is composed of gross matter.

Our analysis then continued to the subtle body. As we saw, the subtle body consists of the inner personality—the mind, the gross and subtle intellects, the memory, and the ego—plus 15 other components. The building blocks of the components of the subtle body are thoughts, which themselves are composed of subtle matter—combinations of pure states of matter.

We then studied the third body, the causal body—the container for the potential gross and subtle bodies during the period when they are unmanifested. Thus, causal matter has the *potential* for gross and subtle matter. As such, it is the subtlest of all forms of matter.

While all three bodies consist of distinct forms of matter, they all share the quality of consisting of *some kind* of matter.

According to both science and common sense, all matter is intrinsically inert and insentient. Look at the gross objects in the world around you—stone, iron, and wood. All these things are inert. The same holds true for all other types of matter, regardless of their subtlety. Thus, it is the intrinsic nature of matter to be inert and insentient.

The gross body, the subtle body, and the causal body—all material entities—are supposed to be inert and insentient. And yet, experientially, we infer that the three bodies are sentient and conscious. This is **because they borrow consciousness from another source. That source is *pure consciousness or original consciousness.* The original consciousness is non-material in nature!**

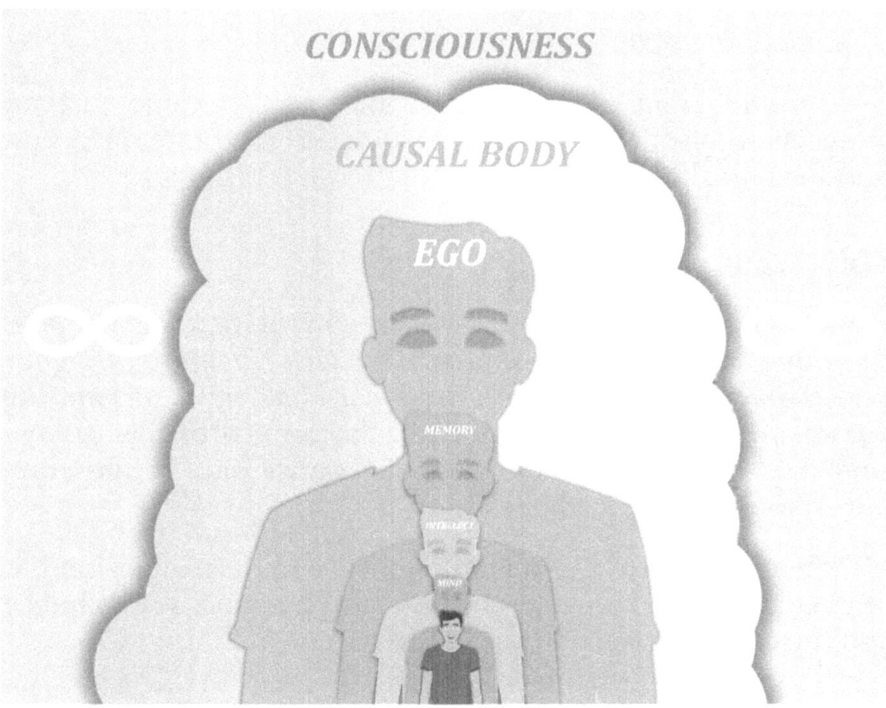

If the original consciousness is also material in nature, then it also must be inert and insentient and would have borrowed its consciousness from another source!

Therefore, the original consciousness is the enlivening principle behind all the three material bodies. They are bathed in original consciousness as it were, to gain sentiency. The original consciousness is another inseparable aspect of your personality.

We will explore the nature of pure consciousness gradually, step by step, as we have done with every aspect of our analysis so far.

Who am I? & Who are YOU?

We observe that an individual is a combination of the borrowed consciousness plus the three bodies combined. What in that combination exactly is you? "Who am I?"

The statement "Who am I?" has now become just an intellectual question, with the least effort to explore deeper to find the real "I" or "You."

YOU - Your belongings

We will begin this exploration with a simple logical statement: ***Anything that belongs to you is not YOU! They are just your possessions, and your possessions are NOT you. You are the possessor.*** What does this mean in the context of this discussion? Let's consider some obvious examples first, before moving on to more nuanced ones.

When you state, "This is ***my*** house, and This is ***my*** car," it implies that the house and car are your possessions. The house and the car belong to you, and they are not YOU!

Likewise, when you state, "This is my hand and this is my body," we need to understand that the hand and the gross physical body belong to you, they are your possessions. Then as a possessor, who are YOU!?

This same logic can also be extended to the subtle body, when you state, "My mind, my intellect, my memory, my ego," it implies that the mind, intellect, memory, and ego are the possessions that belong to you. The subtle body is your possession. Likewise, the causal body too belongs to you. Then as a possessor, who are YOU?!

If you remove all three material bodies and the possessions from the individuality, superficially you may infer that nothing remains. Upon careful rethinking, you will observe a unifying, integrating factor, as subject, a possessor holding the three bodies as possessions. That possessor, the fourth factor in your individuality is the original, pure consciousness. Therefore, the consciousness is the real 'I' or the 'YOU' the possessor of all the three material bodies.

This is one method of deriving the nature of 'YOU.' Let us now consider another.

YOU - Your experiences

Once again, let us begin with a statement: **Anything that you experience is not you.**

The entirety of life, from the moment you were born up to the moment you are reading this sentence, is a constant stream of experiences. As we have indicated previously, experience occurs when a subject and object interact. Is there a way to measure this? Perhaps.

Just as with matter, which can be divided into infinitesimally small fragments, experience can also be similarly divided. That is to say, while you may consider watching a movie an experience, so is watching just one scene or just one frame. **Thus, experience is expressed in terms of the interaction between a subject and an object, regardless of its nature or duration.**

Consider all the experiences that physical and external life consists of—eating, traveling, interacting with people, and so on. Experience requires a subject and an object. When talking about external life, the stimuli associated with activities are objects, while the individual experiencing them is the subject. This implies that the things and objects one experiences are separate from the one who experiences them.

This logic can be extended to the inner personality as well. From the standpoint of the mind as the subject, the body is an object; when the intellect is the subject, the mind is an object. **All the components of the inner personality can be considered as objects. This is both because they are all material in nature *and* because there is one aspect of one's individuality that remains outside of them as a subject and is capable of experiencing them. That aspect, of course, is one's consciousness.**

Even though it may seem like these components of the inner personality are capable of experiencing things, it's really more accurate to say that experience flows through these components. This is because every single one of the components is fundamentally an object, rather than a subject. As such, they can only *appear* to be experiencing something. **The only aspect of one's individuality that cannot be made an object is the consciousness.** Therefore, consciousness is the 'experiencer' within one's individuality.

Let's conclude with a more prosaic example that will encapsulate everything described so far. From the earliest point in life that one can recognize oneself, one entertains a notion of an identity. This identity remains throughout the entire life. "I am a child," is first said; then, "I am a teenager;" then, "I am an adult;" then, "I am an elderly person." Throughout all these stages of life, every part of that individual's mind and body will change—except for their sense of who or what 'I' is. Thus, this 'I' or the consciousness, is the one constant, unchanging phenomenon that exists throughout a person's life.

Chapter 17

Consciousness is YOU!

Exploring Terms

In the previous chapter, our discussion of consciousness was somewhat oblique: we described what it is not (it is not material); we described its basic relationship with matter (it is the medium in which matter is 'bathed,' and from which matter 'borrows' consciousness); and we described its role within the individual (as the ultimate subject, the ultimate 'possessor' and 'someone who experiences').

In this chapter, our discussion will be more pointed. It will focus on the nature of consciousness and will begin to get at the way it fits into the broader conversation around capturing infinity.

The nature of consciousness

To describe the nature of consciousness, we will refer to its *indicators,* rather than using other similar terms like *aspects, qualities,* and so on. In a moment we will see why this distinction is important; for now, just note that the selection of this term is deliberate.

The first indicator of the nature of consciousness is that it is non-material. We know this is the case as matter is inert and insentient; consciousness, then, *cannot possibly* be material, as it is sentience itself. This is its primary indicator.

The second indicator is that consciousness is independent—that is, it is not dependent on anything for its existence or its production. It is self-evident and self-effulgent. It does not originate from the material universe and does not require it for its existence. This fundamental independence is the second indicator of consciousness.

The third indicator of consciousness is that it lacks attributes—it has no attributes, aspects, or qualities. Only material objects can have attributes; from the smallest subatomic particles to the full breadth of the entire observable universe, all matter shares the nature of *possessing* qualities. Since consciousness is non-material and utterly independent of matter, it does not share this nature. This is why we use the term *indicator* to describe its nature. Since it lacks attributes, it does not actually *possess* the qualities that constitute its nature. Instead, they only serve to *indicate* its nature.

The fourth indicator is that consciousness is eternal. The entire observable universe is bound by four dimensions—time being one of the dimensions that belong to the material universe. Since we know that consciousness is not material, we can conclude that it transcends time and is thus eternal. What do we mean by *eternal*? From the perspective of the material universe, it indicates that consciousness exists in all periods of time, has existed since time immemorial and will continue to exist for an unlimited future, even after the Big Crunch or the collapse of the universe. To be more precise, it is utterly distinct from time and cannot be measured by time, because it does not share time's material nature. It simply *is*—permanently.

Let's stay with the scope of the universe for a moment. The universe contains endless space—and yet, it is also material. Because we know that consciousness is non-material, we must conclude that **consciousness also transcends space; this is another indicator of its nature.** Now, does this mean that consciousness only exists *outside* of space? In short, no. It is the nature of matter to exclude other matter; two objects cannot be in the same place at the same time. But consciousness is non-material, which means that it does not share this limitation. **Thus, consciousness not only transcends space but also pervades the entire observable space or universe—another of its indicators.**

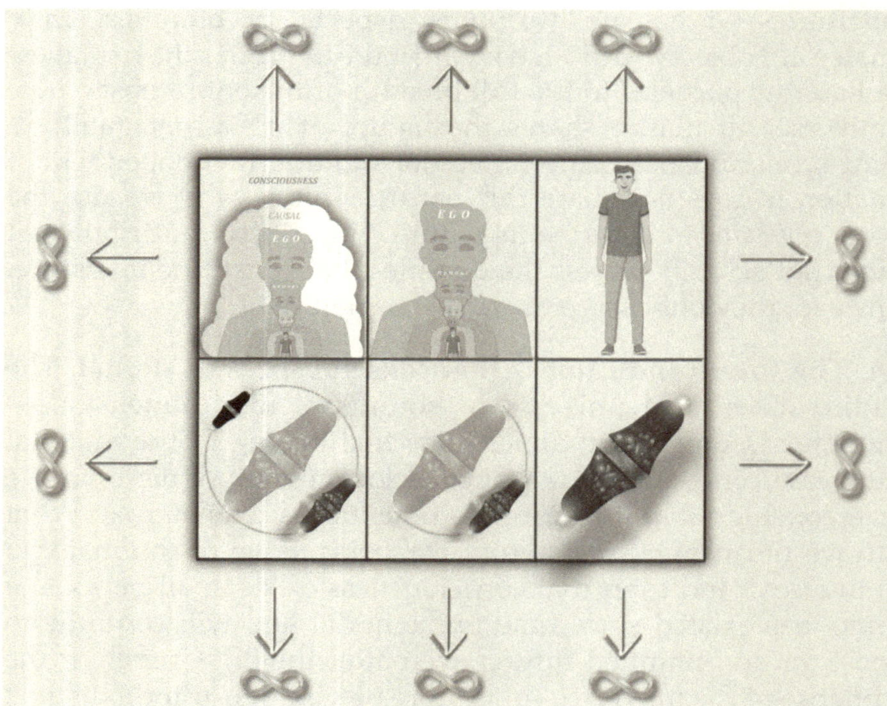

So, consciousness is both eternal and all-pervading. Based on these two indicators, we can ascribe another term to its nature—**consciousness is *immortal*.**

Another indicator of the nature of consciousness is that it is real. What does it mean for something to be *real*? Something is real only when it exists permanently, throughout all periods of time, the past, present and in the future. Anything that changes, ends, or disappears over time is not real. That period can be as brief as an instant, or as long as the period that spans multiple iterations of the manifestation and unmanifestation of the universe. If it ends, it is not real.

We have established that the gross body has a lifespan of a little over 100 years. The subtle body has a lifespan of one iteration of the manifest universe. The universe itself will at some point, be it billions or even trillions of years from now, reach its moment of total dissolution and enter a state of singularity. And even within these fixed lifespans, all matter exists in a state of constant change. This ceaseless, universe-wide state of change indicates that the entirety of the universe is not permanent, and thus cannot be real. Immortal consciousness, however, is not subject to this state of constant change. Therefore, consciousness is the only true reality there is.

The next indicator of consciousness is that it is the knowing principle within an individual. This knowing principle is characterized by the human ability to declare that we know something. We can illustrate this with a simple example. First, we start with a simple statement, "I know your name." Because we can say this and *recognize* that we have said it, we can also make the further statement, "*I know that* I know your name." And we can extend this still further by stating, "*I know that* I know that I know your name." This could go on to infinity. This ability to know and to recognize that we know is the knowing principle—another aspect of consciousness.

Another indicator of consciousness is that it is happiness. This is a somewhat nuanced point, so let's take a moment to understand what it means. We'll start with a few points we've already discussed. We know now that an individual comprises a material component plus consciousness. We also know that consciousness is the ultimate subject within one's individuality. And we know what happiness is—it comes from reducing the number of unfulfilled desires.

Let's start with the last of these points. The unfulfilled desires that accumulate throughout life cause unhappiness because they create a more stressful experience of existence. On the flipside, a sense of happiness accompanies the elimination of unfulfilled desires. Now recall that desires are composed of thoughts and that they are composed of subtle matter. Since matter is inert and insentient, it cannot generate non-material things. Thus, neither the desires themselves nor the objects that satisfy them are the actual source of the happiness we feel when we satisfy desires.

Because of this, we must accept that happiness originates from within, and the only aspect within us capable of originating non-material things is the consciousness. **Therefore, whenever one feels a sense of happiness, it's nothing but a small reflection of the consciousness within one's individuality.** The more desires one eliminates, the more happiness one allows to reflect on their experience of life.

When one manages to negate all material aspects of their experience of life (that is, eliminate all unfulfilled desires), there are no longer any obstacles between one's identity and one's consciousness. In this state, the individual becomes one with their consciousness and thus becomes the embodiment of happiness.

Recall our discussion of the logic behind capturing infinity. We saw that by reducing one's unfulfilled desires to 0, the happiness formula generated an infinite result. **This indicates that eliminating all unfulfilled desires—that is, negating all material aspects of one's experience of life—is how one**

captures infinity. This is another indication of the importance of the relationship between consciousness, happiness, and infinity. Let's continue to explore the precise nature of this relationship.

Together, all the indicators of the nature of consciousness suggest that it is infinite, and is utterly distinct from the finite material universe. As we study the implications of this statement, let's also note how some other key concepts fit in here. In the previous chapter we explored how, when we refer to 'I,' we are really referring to the consciousness aspect of our individuality. Because consciousness is all-pervasive, this means that when we refer to 'you,' we are also referring to the consciousness aspect of another person's individuality. Combined with what we have discussed in this chapter so far, we can state that:

Consciousness = I = You = Infinity = Infinite happiness

Let us spend the remainder of this chapter exploring the implications of this statement.

The implications of infinite consciousness

Essentially our discussions of consciousness and infinity have centered on the meaning of those concepts in the context of human experience. We have focused on the human gross, subtle, and causal bodies, and how consciousness pervades these material forms. Throughout this conversation, though, little attention has been paid to the relationship between consciousness and non-human material forms. Let's touch on that subject now by asking a simple question: **If consciousness, which is responsible for enlivening human beings, is truly all-pervading, what is its relationship with other living beings and inert objects?**

When we say 'other living beings' that term includes all other living things on Earth and other habitable planets of the entire universe, from animals and plants down to the smallest single-celled organisms. The all-pervading consciousness enlivens all the other non-human living beings as well. The reason for their being in this universe will be dealt with in a later chapter on causality.

So, again we ask: how can it be that the all-pervading consciousness enlivens only human beings and other living beings, whereas the inert objects of the universe remain inert without any expression of movement or emotions?

The answer to this far-reaching question is actually rather simple. The **inert objects such as stone/metal/wood, etc. do not express sentience because they do not possess a subtle body.** The subtle body is the instrument that enables inert matter to be enlivened by consciousness. Matter that does not possess a subtle body does not possess the ability to experience the flow of thought. This means that it cannot become sentient. Therefore, consciousness *does,* must and always pervade this inert matter, just as it pervades us; rather than lending it awareness, though, it simply lends this matter the power of *inertia*—a power that permits this matter to exist.

Recall the four powers of the personality layers associated with the subtle body and causal body—acting power, desiring power, knowing power, and inertia power. Human beings possess all these powers. During waking life, the first three are active and define our experience of conscious life; during deep sleep, the inertia power takes over and suppresses the other three. We have seen how, in the state of deep sleep, the personality effectively becomes inert. Well, truly inert forms, which lack the three powers associated with the subtle body, are *always* in this inert state—that is, they *only* possess the inertia power. This power is lent to such matter by consciousness, which pervades all things.

The inertia power is what enables matter to exist at all. All inert matter throughout the entirety of the universe(s) shares this inertia power that underpins their existence itself.

Let's return now to the chain of concepts we linked together at the end of the previous section. We know now that the real 'I' does not refer to the gross, subtle, and causal bodies, but rather to consciousness itself. Thus I = consciousness. We know too that consciousness is non-material in nature, so we can also add that I = consciousness = non-material.

We have also equated both 'I' and 'you' with consciousness and have further described consciousness as the independent entity that enlivens the body, mind, and intellect. It is also both the knowing principle and happiness. However, despite having this nature and conferring these qualities to us, we assert the following: consciousness = without attributes = self-effulgent. In addition to being all-pervading, consciousness is also eternal; it exists everywhere and at all times, always. As such, we can also say that consciousness = eternal and all-pervasive = infinite. And because it is infinite, we can also state that it is non-dual; as an infinite entity, it is simply *one*.

This is what we referred to with the statement consciousness = I = you = infinity = infinite happiness. To this, now, we can also add non-dual or one. We add this to emphasize the fact that not only does consciousness pervade all things, but that all things are united in being pervaded by consciousness. The consciousness in 'I' is the same as the consciousness in 'YOU,' along with the consciousness that exists in everything, everywhere.

Up to now, this conversation has proceeded methodically, moving from one proof to the next. This has been in the interest of establishing a solid, rational foundation for the eventual pursuit and capture of infinity. We have conducted our study in this systematic way both to appeal to readers with backgrounds in the sciences, and also because this method is an effective means of establishing a thorough understanding of the requisite information.

However, there is a whole other realm of thought and study that deals with this pursuit—that of religion. This book has deliberately avoided this area, because it intends to present information rationally, rather than attempting to make dogmatic arguments. However, as we conclude this chapter, let us at least touch on this subject area momentarily. We will not get sidetracked by its logic, but rather show how religious thinking may overlap with the rational and systematic understanding we have laid out up to now. This resonance may serve as another proof for the claim that consciousness is infinite.

There are so many religions in the world, and so many religious figures have come and gone. They all talk about a principle that, for convenience, we will collectively refer to as *God*. This God generally describes some entity or force that pervades the entire universe, enlivening principles and establishing the laws of existence. Though each religion describes this principle differently, they all revolve around the concept of an eternal, all-pervading force—in other words, precisely the definition of consciousness and infinity that we have given via logical analysis.

Now, let us ask a rhetorical question: How many infinities can there be? Of course, there can only be one, as infinity is non-dual. Thus, we can draw the logical conclusion that all these spiritual masters were, in their respective ways, teaching their followers about consciousness, infinity, and how to attain it. Going further, we can also conclude that **if there is some principle that corresponds to God, then it must be infinite, and that GOD must be non-dual or ONE GOD!** In order to explore this premise, let us consider the words of certain religious masters themselves.

In the Hebrew Torah, God reveals to Moses, "I am that I am."

In the Christian New Testament, Jesus Christ says, "The Kingdom of Heaven is within you."

Lord Krishna says, "I am the Self in all beings." And the Vedas declare, "The Self is God."

So, we can infer that different religions invoke the consciousness, infinity, the self, and God in much the same terms.

When we compare statements such as these with the statements of modern scientific thinkers, an interesting pattern emerges. It seems as if ancient religious leaders and modern, cutting-edge scientists wind up using the same language to describe their respective areas of study. Consider the following statements in light of those above, and of the broader discussion we've engaged in throughout this book.

We'll start with quantum physicist Erwin Schrödinger, who said, "Consciousness is never experienced in the plural, only in the singular... of which the plural is unknown... There is only one thing, and that, what seems to be a plurality is merely a series of different aspects of this one thing, produced by a deception."

Max Planck, the originator of quantum theory, stated, "I regard matter as derivative from consciousness. We cannot get behind consciousness. Everything that we talk about, everything that we regard as the existing, postulates consciousness."

Finally, Nobel Prize laureate George Wald said, "Though consciousness is the essential condition for all science, science cannot deal with it... Consciousness itself lies outside the parameters of space and time."

Other scientists, like Einstein and Hawking, have articulated similar views on the subject.

The intention of this brief review is simply to show that, regardless of whether they practice science or religion, individuals who think deeply about this general subject often wind up expressing their conclusions in very similar terms. This lends additional credibility to the claim that there is only one infinity, and that it comprises all the aspects we ascribed to it throughout this chapter and the previous one.

Chapter 18

Big Bang! Who Banged?

Throughout this book, we've indicated that certain topics will be expanded upon in more detail during the discussion of creation. This disclaimer accompanied our analyses of the three bodies, the five layers of the personality, and the three states of experience, along with the previous two chapters dedicated to consciousness.

All those preceding analyses have had value in and of themselves. They can help the reader grow their understanding of their own body, mind and intellect and their relationship to others and the world around them. However, as this book is intended for people belonging to that rare portion of the population interested in and capable of capturing infinity, we will push beyond those earlier subjects and start to connect them with some of the biggest questions humans can contemplate. Why did the universe begin? Where did it come from? What existed before its creation? In short, 'Who banged the Big Bang?' We will begin to explore these all-important questions now as we finally take on the subject of creation.

Cyclical creation

Thanks to the cosmic microwave background and other evidence gained from observations made by advanced astronomical resources up to and including the JWST, we have a thoroughly founded, scientifically validated knowledge of the fact that the Big Bang occurred 13.8 billion years ago.

Advances in technology may lead us to revise this understanding. Some have even claimed that recent observations made by the JWST should already be forcing us to make such revisions, though this opinion represents a minority view within the community. Regardless, all the current evidence suggests that our universe *did manifest* during the Big Bang and that this event occurred at a discernible point some 13.8 billion years ago. Considering that the cosmic cycle spans many iterations of manifestation and dissolution, this event happened relatively recently.

To get back to basics for just a moment—how can we even talk about a 'beginning' to the universe, let alone a cosmic cycle where each universe has its own beginning? It all goes back to the fundamental laws of conservation of matter and energy. Since matter and energy can never be created or destroyed, only transformed, it stands to reason that the entire contents of our present universe could not have come from *nothing*. We have deduced that, 13.8 billion years ago, the entire contents of the universe emerged from the singularity. During this singularity, everything that constitutes our current universe did exist—but only in an unmanifest, dormant, potential state. Now, where did *these* components originate from? What came before them? This is a question that we will analyze soon. But, for now, we must set it aside.

For the time being, so that we have a place to start from, we can simply assert that this unmanifested potential existed and that it predated the formation of our universe. By *unmanifested*, we mean that it existed, but was not perceptible. We have described this state elsewhere as the causal state, and this period of the singularity as the causal universe.

The causal universe is not a subject that has been studied in any great depth by the scientific community. The gross universe has been studied thoroughly, and there have even been some attempts to probe the subtle universe—mainly, by those who probe disconnected aspects of the subtle body; it is not studied as a universe unto itself. The concept of a causal universe, however, remains unexplored and unacknowledged by the scientific community at large.

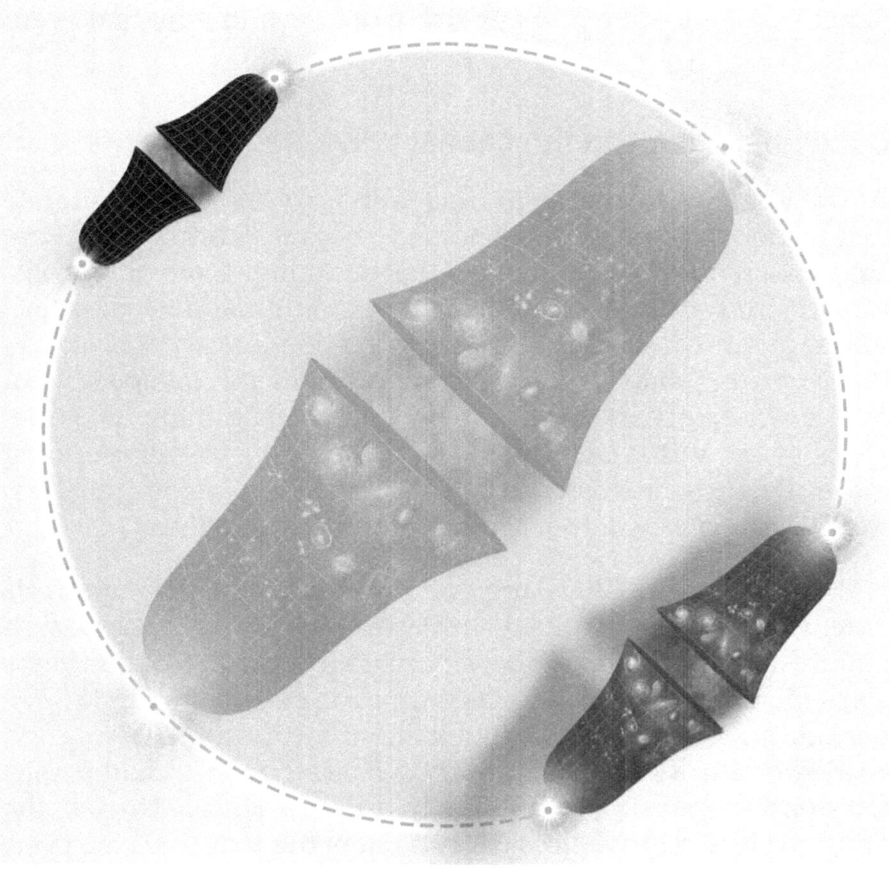

The preceding chapters intended to build a foundation for understanding these nuanced, advanced concepts. We started from the well-understood gross universe and moved toward subtler ideas. Now we are prepared to apply the lessons we learned—specifically what we learned about the microcosmic causal *body*—to the macrocosmic scale of the causal *universe* as we begin to explore the nature of the Big Bang, creation, and what came before it.

We currently inhabit both a gross and a subtle universe, which emerged from the causal universe following the Big Bang. Let's now dive into our study of how and why that event occurred.

Consciousness and the causal universe

As we begin our discussion of creation, we must first reiterate that the term *creation* is a misnomer—because nothing was 'created' when our present universe began; things were only transformed from their potential states in the unmanifested, causal universe. On the cosmic scale, it would be more accurate to say *transformation* than *creation*. However, from the perspective of our present universe, we can consider the Big Bang as a 'creation' event; this is because the contents of the universe sprang into their present existence at that time. This is why the term *creation* is used, even though it is slightly misleading.

So, prior to the Big Bang, the entire universe existed in its potential form—the causal universe. Something triggered its manifestation, and the Big Bang occurred. Some 13.8 billion years elapsed since that time. It will continue to exist, for an unknown period, until it reaches its inevitable point of total dissolution and returns to a state of singularity. That singularity will persist for another period of unknown duration until, finally, the appropriate circumstances trigger a new Big Bang and the cycle begins again, to be repeated endlessly. Throughout this cycle, the causal universe recurs during each period of unmanifestation; therefore, we can observe that the causal universe is an entity, a beginningless entity.

How so? Again, recall the laws of conservation of matter and energy. When the manifest universe reaches its moment of total dissolution, its contents must persist *somehow*. Some scientists have theorized that, when matter and antimatter collide, all matter is annihilated and converted into energy. This, they say, is how the universe's dissolution will occur, a process that will thereafter be reversed with the next Big Bang. Thus, this cycle of manifestation and unmanifestation is beginningless, immortal, eternal, and *infinite* in nature. Still, the question remains: How was this cycle ever set in motion?

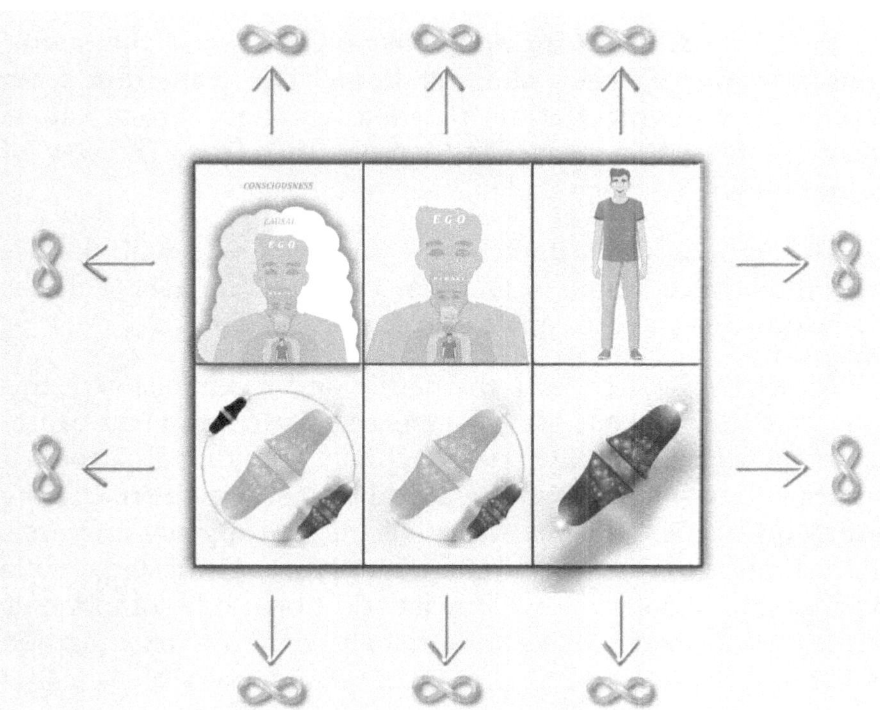

To answer that question, let's compare the beginningless causal universe with another beginningless entity—consciousness itself. Apart from being beginningless, are consciousness and the causal universe alike in other ways? The causal universe comprises causal matter (gross and subtle matter in their potential forms); consciousness is non-material. Because the causal universe is material, we know that it must also be inert and insentient; consciousness, of course, is consciousness in and of itself. The causal universe is continuously changing, locked in its cycle of manifestation and unmanifestation; Consciousness, on the other hand, is eternally changeless. Lastly, as a material entity, the causal universe is finite; consciousness is infinite, pervading throughout the causal universe, bathing it in infinite consciousness—and, thus, lending it sentience.

In the medium of infinite consciousness, it is the law of causation that triggers the Big Bang! The transition from universal unmanifestation to manifestation. From causal universe to subtle universe to gross universe. The law of causation will be discussed in a separate chapter.

Throughout this chapter and the next we will discuss how the interaction between these two beginningless entities, consciousness and the causal universe, occurs.

As we begin to explore the nature and mechanisms of this phenomenon, let us note that the transition from the causal universe to the manifest one is actually not a single event. It occurs in two stages. **First, the causal universe gives rise to the subtle universe, which in turn, gives rise to the gross universe.** In the first stage, 20 material components are manifested (subtle body components), followed by an additional five components in the second stage. Let's start our analysis with an examination of the first stage.

The subtle Big Bang

Throughout this book, we have extensively discussed the nature of the subtle body, but have not treated the subtle universe in as much detail. Therefore, let's begin our examination of the subtle universe on the foundation we've already built, and work toward understanding it using the lessons we've already learned.

We know that the human subtle body consists of the mind, gross and subtle intellects, memory, and ego, in addition to the 15 power centers distributed throughout the body. These 20 components are composed of subtle matter, which itself is composed of combinations of pure states of matter. This, in brief, describes the subtle body of a single human being.

Along with that, let us also collect the subtle bodies of all other human beings on Earth, and add to them any subtle bodies that may belong to comparable life forms existing on habitable planets elsewhere in the universe (if such life forms exist). To this collection, let us also add the subtle bodies of all animal life, terrestrial and extraterrestrial. Animals do possess subtle bodies, though they lack the intellectual components. Plants, too, despite being primarily gross in nature, have trace amounts of subtle bodies. Add these subtle bodies also to our collection, along with those that originate on Earth and those that exist throughout the cosmos.

Inert objects do not possess subtle bodies. However, we did establish that they are pervaded by a power of inertia that is responsible for their ability to exist at all. While this power is associated with the causal body, recall that the hierarchy of subtlety holds that subtler material pervades grosser material. This means that the power of inertia pervades throughout the subtle universe. Therefore, this is a kind of subtle power that pervades inert matter, making it a part of the subtle universe as well.

Now our collection consists of all subtle bodies, from that of a single human being on planet Earth to the last microscopic organism at the furthest reaches of the observable universe, in addition to the subtle power that pervades all inert matter, everywhere. And since what we refer to as the 'observable' universe is nothing but an artificial distinction generated by mankind's own limitations, we can conclude that this subtlety also pervades throughout the entirety of the unobservable universe, and throughout any other manifest universes that may exist beyond the boundaries of our present one.

This aggregation of all subtle bodies and all subtle power throughout all manifest universes represents what we refer to as the subtle universe.

The lack of study into the nature of the subtle universe represents a huge gap in the scientific community's understanding of creation. When we speak of the Big Bang, most analysis proceeds directly from the singularity (that is, the causal universe) to the gross physical universe; the subtle universe is entirely neglected. Yet the creation of the subtle universe represents a distinct stage of creation unto itself. **During the manifest universe's creation event, triggered by the law of causation, the causal universe first gives way to a subtle Big Bang, which then gives way to the gross Big Bang.** To ignore the subtle universe is not only to remain ignorant of *its* nature, but of the nature of creation, the singularity, and the gross universe as well. With that in mind, let's examine how the subtle Big Bang occurs, to fill in this gap in the general scientific knowledge on the subject of creation.

The three intrinsic properties of the universe

We'll begin our analysis of the subtle universe with a study of the creation of its constituent parts—that is, of subtle matter itself. Again, the term *creation* here is a misnomer, since in reality this matter is not being 'created' but rather transformed. From the perspective of this manifest universe, though, the event can be viewed as a kind of creation, so we will use the common (though admittedly inaccurate) term *creation*. However, before getting to the subject of subtle matter, we must first introduce an additional concept that has a significant bearing on the universe and its creation. This concept describes the three intrinsic properties of the universe.

- ***INERTIA COMPONENTS***
- ***ACTION COMPONENTS***
- ***KNOWLEDGE COMPONENTS***

Once again, the quickest way to establish a foundation for this concept is by beginning with familiar terrestrial examples and then building our understanding from there. So let's begin by observing human behavior. Throughout an individual's life, all of their behavior will fall into one of three categories. First, they will take in and interact with knowledge; this is a continuous process of ***knowing*** that unfolds throughout one's waking life. Then, there is ***acting***, which also occurs throughout waking life. Based on the outcomes of one's process of knowing, one will commit to performing actions in the physical world. Finally, there is ***resting***. With all this knowing and acting, the individual's gross and subtle faculties tire over time and require rest. Thus, the entirety of human experience comprises these three categories of behavior: knowing, acting, and resting—or expressed another way, *knowledge, action, and inertia.*

Let's compare the experience of humans with that of animals. Animals may have the ability to conduct a rudimentary form of knowing, but, because they lack intellect, they are not able to perform this behavior in the same manner as humans. However, they are clearly capable of both acting and resting; thus, their experience of life comprises those two categories, with slight traces of knowing.

Plants, in turn, are capable of exerting some rudimentary forms of action—they conduct photosynthesis, i.e. process carbon dioxide,and release oxygen. But their predominant form of behavior is resting. This single category of behavior characterizes their existence, with a slight trace of acting.

So much for living things—what about inert matter? Obviously, inert matter neither knows nor acts. It simply exists, driven by the power of inertia. Thus, the existence of inert objects is characterized solely by the behavior of resting or inertia.

Therefore, we can observe that the entire contents of the manifest universe are characterized by the combination of the properties of knowledge, action, and inertia—its intrinsic properties.

To connect this to the broader conversation, consider what we already know about the relationship between the unmanifested, causal, and manifested universes. We know that everything that exists within the manifest universe exists within the unmanifested one in its potential form. **So the causal universe must contain the potential forms of these three intrinsic properties, along with all the other potential.** Throughout the remainder of this chapter and in the next one, we will further develop our understanding of the relationship between these three intrinsic properties and subtle matter, as well as their significance in the context of creation.

The creation of subtle matter

Because the universe is nothing but its constituent parts, when we talk about creation we're really talking about the creation of those parts. The constituent of the subtle universe is subtle matter. We discussed the nature of subtle matter at length in the chapter on the raw materials of the subtle body. Let's briefly reprise that discussion to reassert our understanding of the concept in light of the present conversation.

Recall that, in the gross material universe, there are numerous states of matter—research has identified more than 20 states of matter, with more likely to be identified in time. That said, gross matter nonetheless boils down to five fundamental states—space, gas, plasma, liquid, and solid. Of these, space tends to be the most difficult one to apprehend, or even accept, as a class of matter. However, general relativity has revealed that space is an aspect of the gravitational field of the universe, and phenomena such as gravitational lensing have demonstrated that space is, indeed, matter. Thus, we can consider these five states as the fundamental states of gross matter.

In our previous analysis, we saw how the subtle universe is composed of these same five fundamental states of matter in their pure form. We can refer to them as the five pure fundamental states of matter. *How these five pure fundamental states of matter are generated from their potential forms during the Big Bang will be the central focus of our present study.*

In the same way that we have adopted the term *creation* on account of its general usage and relevance from the perspective of our gross universe, we can also refer to these pure states of fundamental matter by their more common equivalents. This is done to enable easier visualization and regardless of the term used, the meaning remains the same. Thus, we will continue to refer to space as *space* but will use *air* in place of *gas, fire* for *plasma, water* will replace *liquid,* and *earth* will represent *solid.*

Now, as we saw previously, for a given state of matter to be considered pure, it must contain no trace of any of the other states of matter. It must be itself and only itself. For example, for a given quantity of water to be pure, it must comprise water alone, with no trace of space, air, fire, or earth. Let's revisit a simple, tangible example of this so that we can reestablish the logic behind pure matter. With that logic, we can extrapolate the nature of the entirety of the subtle universe—and, in doing so, the *creation* of the subtle universe from the preceding state of being unmanifested.

Recall our example of water. In the gross universe, this water will inevitably consist of some amount of each of the four other states of gross matter. It will contain some trace minerals and thus some degree of earth. It will possess some temperature and thus some degree of fire. It will also be permeated by some amount of air, and by mere virtue of existing in the dimensional space, will contain space as well. Even the purest of the pure water will contain something more. It is the nature of gross matter to be imbued with some degree of all of the other states of matter.

The reader will recall from our previous discussion that matter can only exist in its pure state—excluding all four of the other fundamental states of matter—in the form of thought. Let's take the time to recall the exact nature of a pure state of matter because this point is a foundational one for our analysis going forward.

First, if matter in its gross, adulterated state is material in nature, then matter in its pure state must also be material. In other words, matter does not cease to be matter when it is purified. Next, if matter in the gross physical universe is inevitably impure, it follows that pure matter cannot exist there. However, we know that it must exist *somewhere*. Thus, the only place it can exist is in the subtle universe, and the only way it can exist is in the form of thought.

Now, what precisely does this mean? Consider again that quantity of water—but, rather than imagining gross water, consider *pure* water, without any of the other states of matter in it. This is impossible in the gross world, but entirely possible within the mind.

Next, try removing that precise amount of pure water from the thought of it; when this is done, nothing remains. Put another way, the pure state of water only exists when the thought of it does, and the thought vanishes when the pure state of water is removed. **This indicates that pure states of matter are identical with thought itself.** The *reader should take a few moments to dwell on that statement* because it is critical to our study going forward.

We have just deduced how thoughts are composed of pure states of matter. However, thought—and the entire nature of the subtle body—consists of much more than just thoughts of pure states of matter. The flow of thought inside an individual contains their memories, their emotions, their active choices, their unfiltered desires, and their physiological mandates. It is complex, constantly changing, and essentially limitless. Thus, **specific thoughts are composed of *combinations* of the five pure fundamental states of matter.**

We can observe this in principle with a brief thought exercise. Take a moment to let your mind wander. Inevitably, several complex thoughts will crop up. Now take those thoughts and consider them as a whole, they may be of some objects or some beings. From that whole, try to systematically remove all the pure states of matter. In doing so, you will gradually observe that nothing of the original pool of thought remains. This is a simple means of observing that even complex thoughts consist of combinations of pure states of matter and that, without such matter, the thoughts themselves cannot exist.

To sum up: **Subtle matter is matter. It is composed of the five pure states of fundamental matter. They are the building blocks of thought itself and complex thoughts are composed of combinations of the five pure states of fundamental matter.**

The intrinsic properties of subtle matter

With a sound understanding of the nature of subtle matter in place, let's bring in the concept of the intrinsic properties of the universe. We indicated that the manifest universe is characterized by these three properties and that they must exist in the causal universe in their potential forms as well. Because the causal universe comprises nothing but subtle and gross matter in the potential states, it follows that any aspect that pertains to causal matter *also* pertains to the potential of other states as well. **Therefore, the five pure states of fundamental matter are also imbued with these three intrinsic properties—knowledge, action, and inertia.** These properties do not adulterate the purity of the various states of subtle matter, they are simply ingrained aspects of them and are constituent parts of their purity.

The image illustrates this principle. Each of the five pure states of matter—space, air, fire, water, and earth—all contain the properties of knowledge, action, and inertia as fundamental components.

Now, what role do these properties play in the creation of the universe? Beyond that, what role do they play in the broader cycle of manifestation and unmanifestation? We can answer this question by invoking Newton's first law: An object at rest remains at rest, or if in motion, remains in motion at a constant velocity unless acted upon by a net external force.

We have associated the causal universe or singularity with the quality of inertia. It contains all the matter of the gross and subtle universes, but in potential form only. Thus, we can consider the causal universe as an object at rest. This implies that the manifest universe represents the same object in motion. This begs the question, 'What external force acted upon the causal universe in order to set it in motion?'

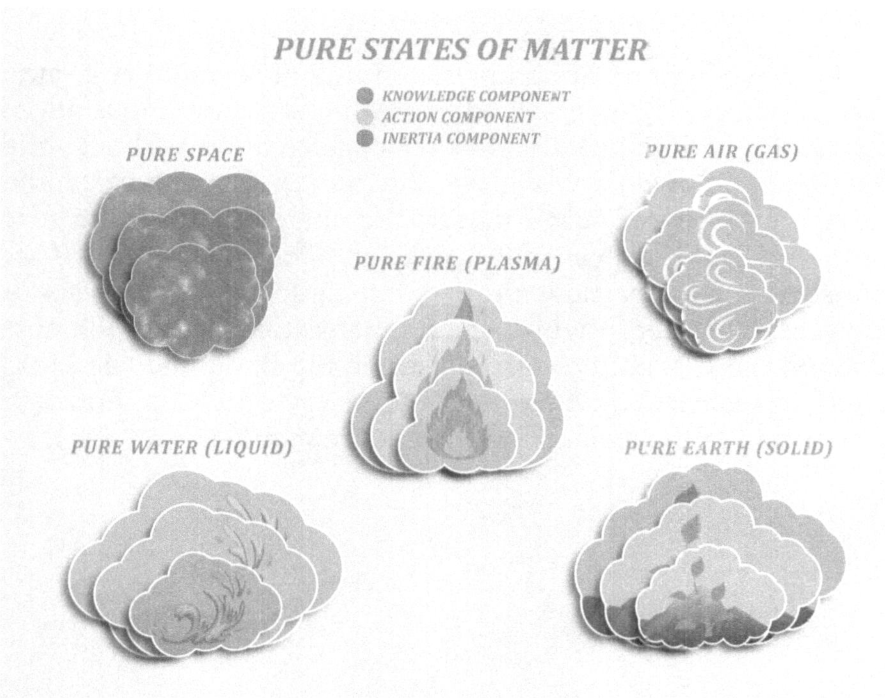

The external force that acted upon the causal universe and triggered the Big Bang is nothing less than the cumulative effect of all the past actions of all living beings throughout the

entire cosmos or 'The Law of Causation.' This effect is the force that disturbs the equilibrium and triggers the transformation of the potential, unmanifested universe into the manifest one, in the medium of consciousness. We will soon study the nature of this effect, described in terms of *causality principles and the law of causation*. These principles are identical to the concept of karma. However, both because that term has been excessively commercialized and because this book intends to present its material objectively and scientifically, we will use the term *causality principles* instead. We will study these principles in detail, including the specific manner in which they exert force that leads to both the manifestation *and the unmanifestation* of the universe, in a later chapter.

We have built a strong understanding of several key topics: the cyclicality of universal manifestation and unmanifestation; the nature of manifestation (a two-stage creation with a subtle Big Bang followed by a gross Big Bang); the nature of the subtle universe and subtle matter; the universe's three intrinsic properties; and the causality principles that trigger universal transitions. We are now prepared to launch our analysis of the actual manner in which the Big Bang unfolded. In the next chapter, we will study both the subtle Big Bang and the gross Big Bang—and, in doing so, will ultimately provide a thorough accounting of the formation of the entire manifest universe.

Chapter 19

Unraveling the Big Bangs

At the end of the previous chapter, we invoked Newton's first law to demonstrate the scientific basis for the causality principles (otherwise known as the law of causation or karma) and the cyclicality of manifestation and un-manifestation. In this chapter, we'll invoke Einstein's mass-energy equivalence formula for the same purpose.

Einstein's famous formula $E=mc^2$ articulates the relationship between matter and energy. These variables differ only in terms of their units of measurement and the multiplicative constant of the speed of light. This indicates that comparatively small amounts of matter correspond to vast amounts of energy, and vice versa. This principle may account for the translation of matter to energy during the total dissolution of the manifest universe—perhaps, as some have indicated, as a result of the obliteration of matter and antimatter. Given that the entire mass of the universe would turn into energy during this event, the amount of energy condensed into the unmanifest, causal universe would be staggering.

This overwhelming amount of energy would then be available to be released and translated into matter during the subsequent Big Bang event, triggered by causality principles. Thus, the mass-energy equivalence matches the claim that the state of the universe is cyclical in nature, cycling between a potential, unmanifested state, and a material, manifested one.

This statement does not deviate much from the consensus among the scientific community. There is, of course, a major difference between the theory most scientific thinkers use to explain the event of creation and the explanation of the same event that has been laid out in this book. Specifically, the consensus holds that the Big Bang was a single event, during which an essentially gross universe sprang from a gross singularity. This book has presented a different case—that the Big Bang actually comprises two separate events: the derivation of the subtle universe from the causal one, and the derivation of the gross universe from the subtle one.

The general lack of scientific study of, or even knowledge of, the subtle universe and the subtle Big Bang represents a major hindrance to mankind's ability to understand the nuances of the manifest universe. Contributing to this subject is one of the primary goals of this book. The hope is that increased awareness and consideration of the nature and genesis of the subtle universe may lead scientific thinkers to new insights in their fields, and even toward the elusive 'Grand Theory of Everything'—a development that would undoubtedly benefit all of mankind.

With that said, let's commence our analysis of the two Big Bang events, starting with the subtle Big Bang and discussing in detail how these transitions unfold.

How the subtle Big Bang unfolds

When causality principles exert their destabilizing force on the causal universe and trigger the release of its unthinkably vast potential, the result is the subtle Big Bang. **This event is first characterized by the formation of subtle matter and then by**

the formation of other subtle entities. These formative events occur in a specific order, which we will outline now.

The formation of subtle matter

During the subtle Big Bang, subtle matter manifests as the five pure states of fundamental matter. **These states manifest in a specific order, based on their degree of subtlety.**

Thus, space is the first of the pure states of fundamental matter to manifest. Space is followed by air (gas), fire (plasma), water (liquid), and finally earth (solid).

What determines the order in which these states of matter manifest? Recall the degrees of subtlety assigned to these states during our discussion of the components and functions of the subtle body. Depending on the perceptibility of each state of matter, it is considered more subtle or less subtle, with the least perceptible state being the most subtle. During the subtle Big Bang, the states of matter manifested in order from most rarified to the least.

As pure space is the most rarified state of matter (associated with only one sense property—sound), it is the first to manifest. This is not to say that pure space *itself* can be heard, or that sound *itself* is also manifested during this stage. Instead, sound is only to be understood as a sensory property of space, and since it is space's *only* sensory property, it is considered the most rarified state of matter.

Pure air is the next state to manifest, associated with both sound and touch. Next is pure fire, associated with sound, touch, and sight. This is followed by pure water—associated with sound, touch, sight, and taste—and, lastly, by pure earth—associated with all the senses.

Each of these states of matter emerges from the one that precedes it. Pure space emerges directly from the causal universe, but pure air emerges from pure space, pure fire from pure air, and so on. Also note that, as indicated in the graphic above, each pure state of matter is associated with one unique sensory property of its own, in addition to all properties belonging to the state of matter from which it originated. For example, pure space is solely associated with its own unique property—sound—while pure air is associated with sound in addition to its unique property—touch. This continues to pure earth, which is associated with the other four sensory properties, plus its own unique property—smell. These inherited, non-unique properties are known as *parental properties*.

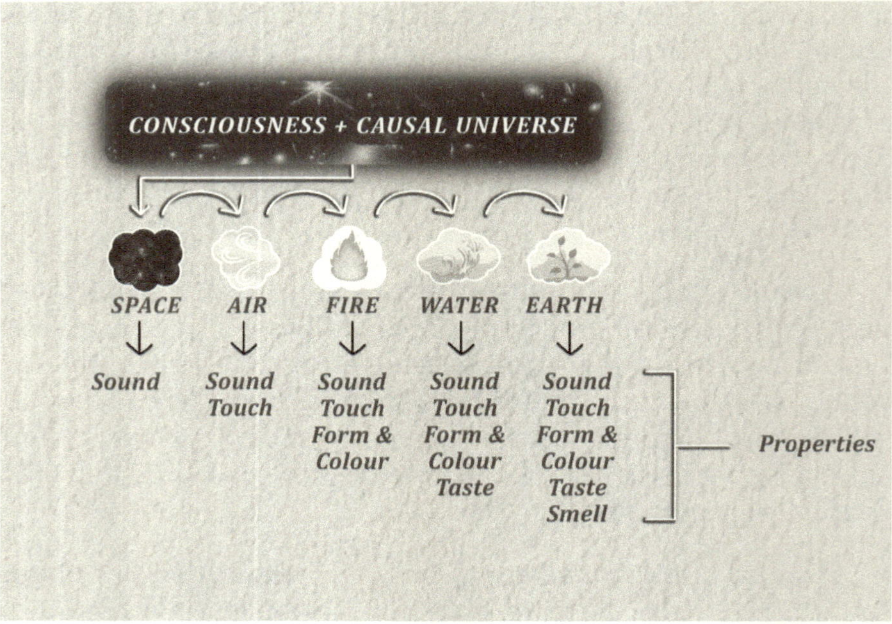

Once more, the reader should note that these senses don't actually exist at this time. Indeed, there isn't even anything *to* be sensed in the universe yet. The universe is only engaged in the formation of subtle matter, which is pure thought. This alone exists at this time, and nothing more.

Beyond their sensory properties, each pure state of matter also contains another set of properties—the intrinsic properties we introduced in the previous chapter. Unlike the sensory properties, which vary in expression from one state of matter to the next, **these three intrinsic properties—knowledge, action, and inertia—are equally ingrained in all five pure states of fundamental matter**, winding through them almost like a DNA strand through the cosmos.

With the manifestation of pure earth, the fifth pure state of matter, the phase of the subtle Big Bang that involves the formation of all the raw materials of the subtle universe on the cosmic scale is complete. This phase is followed by the manifestation of various other subtle components, which will also manifest in a specific order based on their nature.

The formation of the subtle body: Knowledge

The subtle body is the next subtle entity to form after the manifestation of the five pure states of fundamental matter. This involves the manifestation of the 20 components of the subtle body, which form in a specific sequence. **This happens in two phases, the first of which is driven by the intrinsic knowledge embedded within the pure states of matter. The second is driven by the intrinsic action.**

This first phase includes the formation of 10 subtle components—the five power centers of perception and the five components of the inner personality. The five power centers of perception form first, in the same order as the pure states of matter associated with the sense to which they correspond.

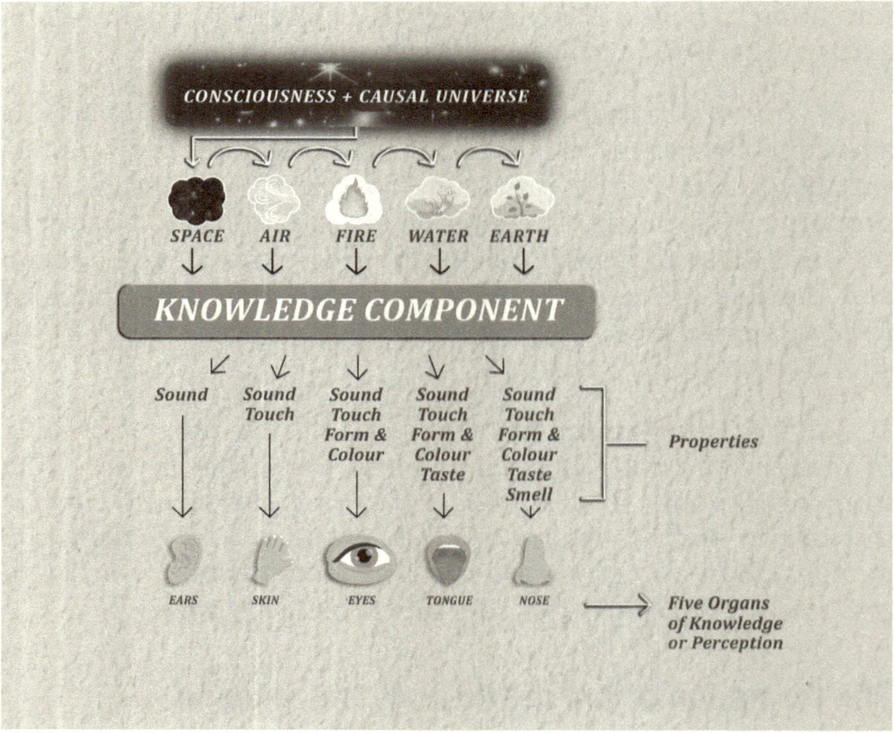

This means that the power center of hearing forms first, followed by the power centers of touch, sight, taste, and lastly smell. These five power centers of perception can also be referred to as the *subtle organs of knowledge* since they derive from the intrinsic property of knowledge.

These subtle organs of knowledge are precipitated by the intrinsic property of knowledge within each of the five pure states of matter. Beyond that, each of these subtle organs derives directly from the specific state of matter with which it shares an association. **Therefore, it is the combination of intrinsic knowledge with the specific sensory association of each pure state of matter that is directly responsible for the formation of each of these components.** Let's briefly run through each component to see what this means.

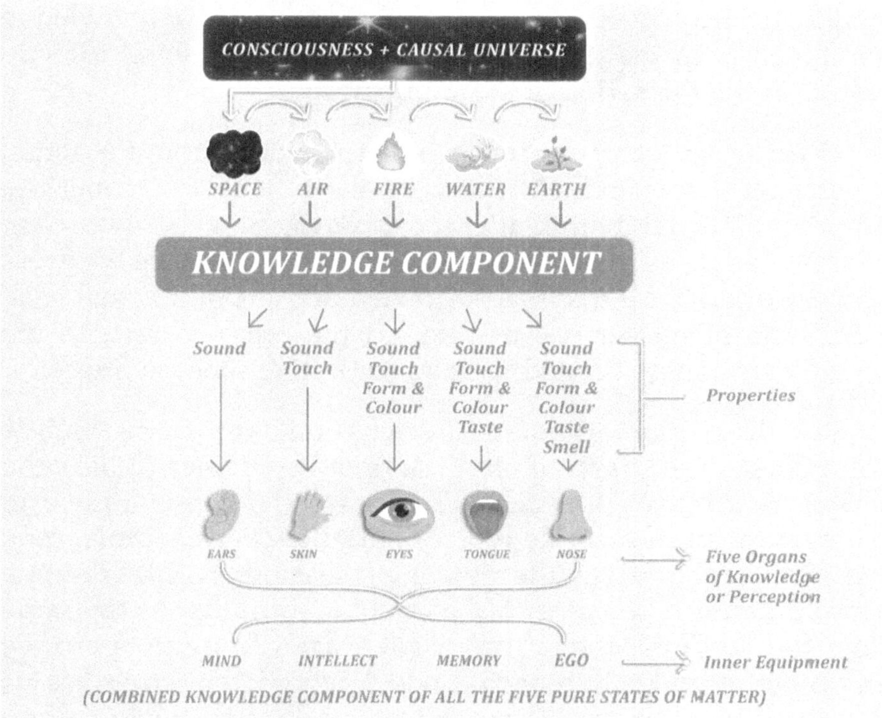

From pure space, the intrinsic property of knowledge combined with the unique sensory property of sound directly results in the formation of the subtle ear.

From pure air, the intrinsic property of knowledge combined with the unique sensory property of touch directly results in the formation of the subtle skin.

From pure fire, the intrinsic property of knowledge combined with the unique sensory property of sight directly results in the formation of the subtle eye.

From pure water, the intrinsic property of knowledge combined with the unique sensory property of taste directly results in the formation of the subtle tongue.

Finally, from pure earth, the intrinsic property of knowledge combined with the unique sensory property of smell directly results in the formation of the subtle nose.

Once again, we are not even close to talking about the actual sensations associated with any of these sensory properties. These don't develop until after the gross Big Bang has occurred, creating the gross matter that will eventually compose the gross sensory organs. All that has happened at this point is that each pure state of matter has manifested the *power to perceive* the sensory stimuli associated with its respective sensory property.

Recalling our previous discussion of these subtle components (not to mention one's own experience of life), the power associated with each subtle sensory organ is limited to the perception of a single type of stimulus. For example, only the subtle ear enables the power of hearing; no other subtle sense organ can perceive that particular stimulus. At the same time, the subtle ear is *only* capable of hearing; it cannot perceive any other stimulus except that one. Each power center has its dedicated function that cannot be interchanged with any other.

The subtle components that enable coordination between these power centers of perception are the next to form. They are, of course, the components of the inner personality—the mind, subtle and gross intellects, memory, and ego. These five subtle components function as intermediaries between the various power centers, enabling a synthetic perception of matter—or, put another way, of *combinations* of matter. This synthetic power is what enables the pure flow of thought from the five pure states of matter to become unique, discernible *thoughts.*

Unlike the five power centers, which are derived directly from single, pure states of matter, **the five components of the inner personality are generated by the combination of the knowledge within *all five* pure states of matter.**

So these first 10 components of the subtle body are generated from various combinations of the five pure states of matter with the intrinsic knowledge contained within that matter. With the formation of these 10 components, the knowledge property expends itself and is not involved in any further creation during the subtle Big Bang. The action property is responsible for the creation of the remaining 10 components of the subtle body; let's explore this next.

The formation of the subtle body: Action

The next subtle components to manifest are the power centers of action—speech, manipulation, locomotion, elimination, and reproduction. Again, at this point, we are not talking about the organs or systems that facilitate these actions, but only about the subtle powers that furnish the capacity to carry them out.

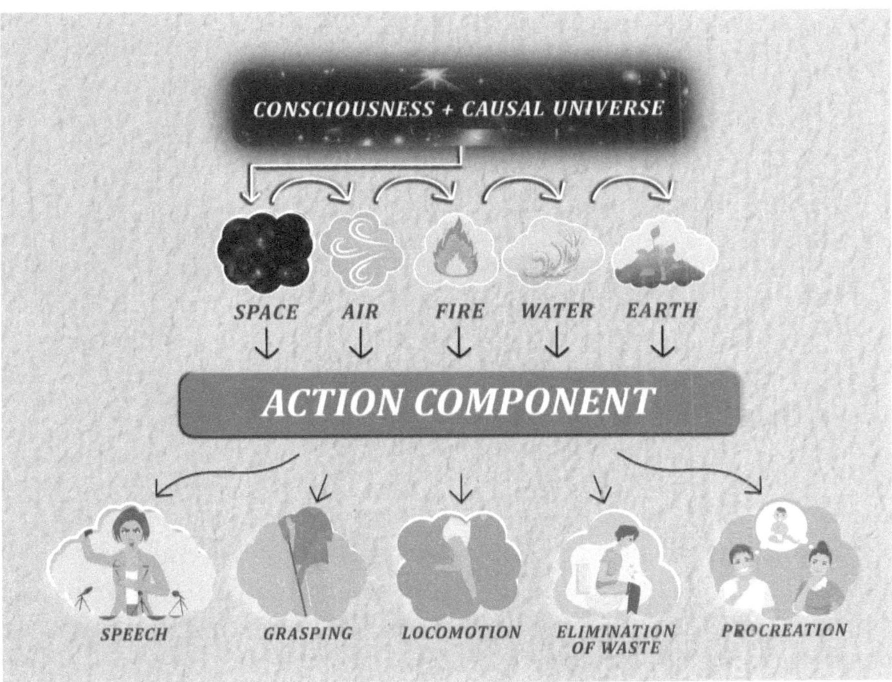

Just as in the previous phase, these power centers derive directly from each pure state of matter as they exert one of their intrinsic properties. In this case, however, they are exerting their intrinsic property of *action*. The power centers of speech, manipulation, locomotion, elimination, and reproduction are respectively associated with the pure states of space, air, fire, water, and earth. They manifest directly from them, via the exertion of their intrinsic action property, as indicated below.

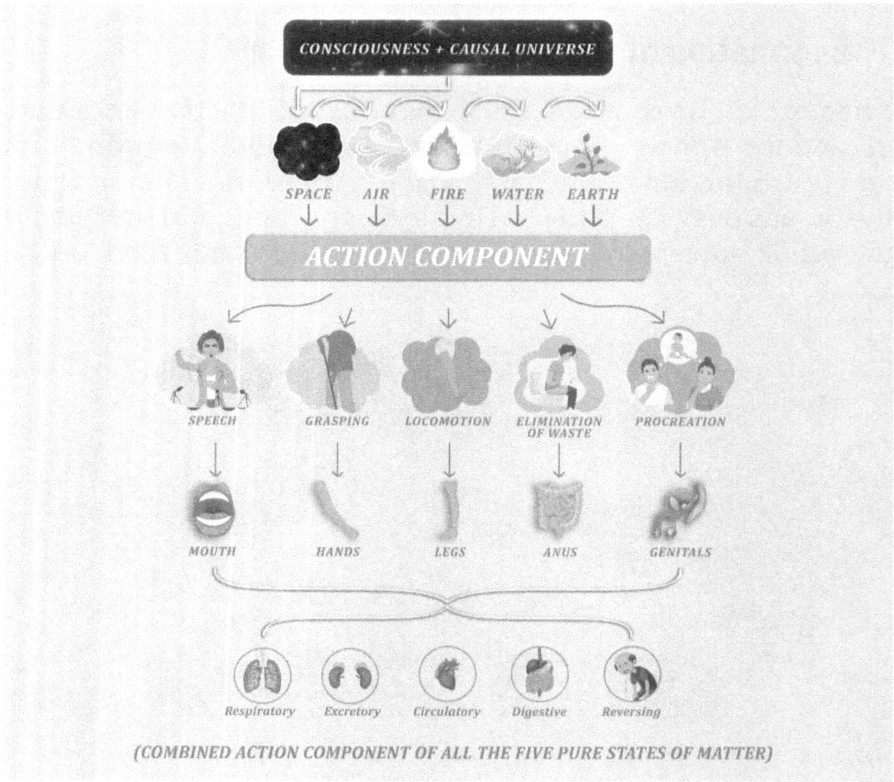

Similar to the power centers of perception, the power center associated with each type of action is limited to that particular action alone. The power center of speech is the only one capable of performing speech, and is incapable of performing any other action. Like earlier, each power center has its own dedicated function that cannot be interchanged with any other.

The subtle components that enable coordination between the power centers of action are the next components to form. They are the five power centers of physiological function—respiration, digestion, circulation, excretion, and reversal. The functions of these five power centers are intertwined. Together, they combine to allow the remainder of the subtle components to operate.

Thus, these five components are synthetic in nature, rather than specific. This is because, just as we saw in the previous section, **the five power centers of physiological function are generated by the combination of the action property within *all five* pure states of matter**, as opposed to being derived from individual ones.

With the creation of these 10 subtle components, the intrinsic property of *action* has effectively expended itself and will make no further contributions during the subtle Big Bang. In fact, **following the expenditure of this intrinsic property and the generation of all 20 subtle components, the subtle Big Bang is complete, and the gross Big Bang ensues.** However, before moving on, let's take a moment to look beyond the context of the human body and consider the implications of the subtle Big Bang for other entities.

Subtle creation beyond the context of the human subtle body

We will consider animal life first. Though the precise configuration varies slightly from one form of life to another, the animal's subtle body generally consists of 17 subtle components. These subtle components correspond to the human subtle components, except for the ego and the gross and subtle intellects.

Some animals may have fewer than 17 components. They may lack a certain power center of perception or action, or may even lack memory, but 17 represents the *maximum* number of components present in an animal's subtle body. The subtle bodies of animals are manifested concurrently with the manifestation of the subtle bodies of humans, and the processes unfold in the same manner.

The subtle bodies of plants are different. Unlike humans and animals, which possess numerous distinct subtle components, plants possess a set of amalgamated subtle components. The intrinsic property of knowledge within the five pure states of matter generates an amalgamated perceptive subtle component; similarly, the intrinsic property of action generates active and energy components (the plant equivalent of the five power centers of physiological function). Thus, the subtle bodies of plants comprise just three subtle components. They are formed during the phase of the subtle Big Bang that pertains to the intrinsic property from which they derive.

Finally, what about inert matter? Inert matter does not possess any manner of subtle body, so no phase of subtle creation pertains to it. However, recall that we previously identified a subtle power of inertia that is responsible for enabling all matter to exist at all. This power is distinct from the *intrinsic property* of inertia, which asserts itself during the gross Big Bang. This is the causally derived power that permeates all subtle matter that simply allows it to exist. As such, it is present in all matter right from the beginning of the subtle Big Bang. So, inert objects possess this subtle power though they do not possess a subtle body.

From these abstract descriptions of the various subtle bodies and subtle components, we can extrapolate the manifestation of *all* subtle bodies and *all* subtle components for *all* the contents of the entire manifest universe(s). *All* the five pure states of fundamental matter, *all* the subtle components of human and animal life, *all* the subtle amalgamations of plant life, and the subtle power of inertia that permeates all inert objects—together form the subtle universe, which manifests directly from the causal universe.

At this point in the universal manifestation, not one atom of gross matter has yet been formed. The state of the universe consists of pure thought alone (which travels faster than the speed of light). However, with the expenditure of the intrinsic properties of knowledge and action, the manifestation of the subtle universe is complete. The subtle Big Bang has concluded, and the gross Big Bang is set to begin!

How the gross Big Bang unfolds

During the subtle Big Bang, the intrinsic properties of *knowledge* and *action* expended themselves to manifest the subtle universe. However, the third intrinsic property—*inertia*—remains. **The expenditure of the intrinsic property of *inertia* is responsible for triggering the gross Big Bang and manifesting all gross matter of the gross universe.**

For many people, *this* event represents the formative event of the universe—'the' Big Bang. This perspective does not consider the formation of the subtle universe at all. Even those engaged in the contemplation of various aspects of the subtle universe are not fully cognizant of the subtle universe *itself*—never mind the subtle Big Bang. Of course, without an understanding of these subtle phenomena, one's ability to truly understand the nature of the universe will be limited. This is why it's so important to fill this gap in the general understanding of universal manifestation. Our attempt to do so continues in this section, as we explore how the subtle Big Bang gives way to the gross one.

Recall the example of water. The quantity of water could exist in its pure state only in thought form. In its gross state, it inherently contains some portion of the other four states of matter as well. Thus, the inherent characteristic of gross matter is that it is a combination of *all five states* of gross matter. To understand the logic behind this characteristic, let's examine how gross matter is derived from subtle matter.

Once the five pure states of fundamental matter expend their intrinsic properties of knowledge and action, only the property of inertia remains; and it now asserts itself. ***The exertion of the intrinsic property of inertia represents the start of the gross Big Bang.***

Each pure state of matter exerts the intrinsic property of inertia within it, manifesting each state of gross matter. However, as we have established, each state of gross matter inherently contains all other states too. **Thus, as pure space manifests gross space, pure air manifests gross air, and so on, each of these pure states of matter instantly combines with all other four pure states in specific proportions, creating the *true* form of each specific gross state of matter.** How the various states of gross matter manifest and combine is illustrated in the following chart.

During the subtle Big Bang, the subtle matter was distributed throughout the universe(s). Thus, when the inertia property exerts itself, gross matter is deposited throughout the universe(s) as well. Put a slightly different way, because we know that each state of gross matter contains some portion of the other four states, wherever *any* one state of gross matter exists, the other four states will exist within it, in equal proportions.

For example, gross water will contain 50% pure water and the other 50% will be occupied by the remaining four states of matter, namely space, air, fire, and earth, say 12.5% each.

This percentage is distributed on a galactic, cosmic scale, not confined just to planet Earth.

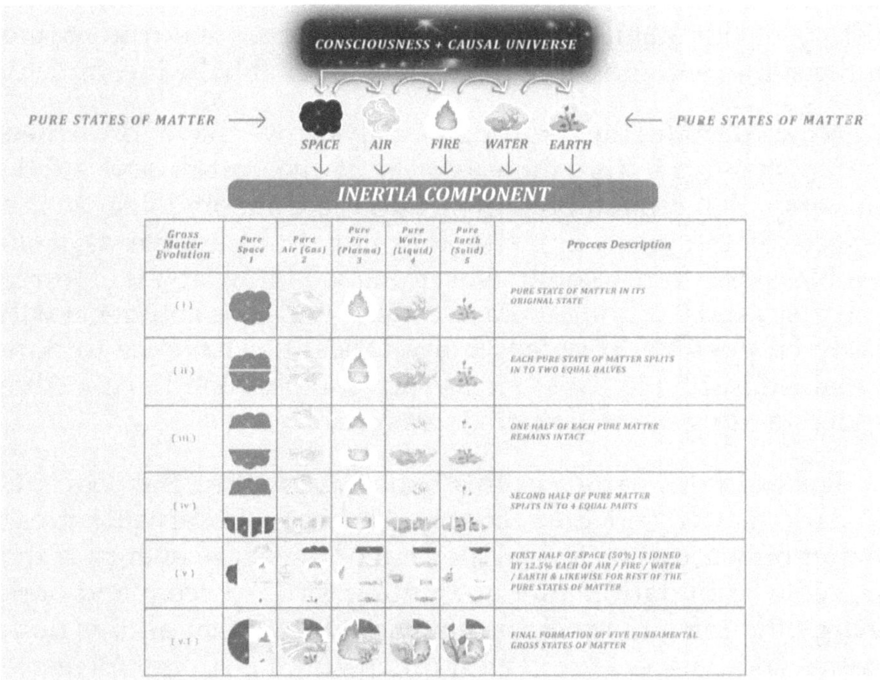

In this manner, the gross Big Bang derives directly from the subtle Big Bang and results in the formation of the five states of gross matter via the exertion of the intrinsic property of inertia. **Once this has occurred, the 'Big Bang' (as it is commonly understood) effectively begins**—from the entrenchment of the physical laws of the universe to the manifestation of the first elements of the periodic table; from subtle agglomerations of trace elements in space to the formation of stars, planets, galaxies, and all other astronomical bodies; and from the origin of microscopic life forms to the arrival of mankind. The nature and sequence of each of these processes have been well documented by scientific observers and form the so-called history of our universe. Of course, we have just seen how this history did not even commence until after the *true* origins of the manifest universe had occurred. As such, any accounting of the history of the manifest universe that excludes the true nature of the subtle and gross Big Bangs will inevitably be incomplete.

Now, despite the thorough analysis we have presented so far, there are still some aspects of the causal and subtle universes that remain unknown; and there are obviously many aspects of the gross universe that remain unknown as well. But by taking into account everything we know about all three universes and studying them in concert with one another, it will likely be possible to untangle mysteries that have up to now remained insoluble. This can also advance mankind's collective understanding of the universes we inhabit.

For example, earlier in this book we invoked the concepts of dark matter and dark energy. Ordinary, observable gross matter only accounts for five percent of the entire energy of the universe, with dark matter accounting for 27 percent and dark energy the remaining 68 percent. Could it be that all this dark matter and dark energy corresponds with the subtle universe? Could it specifically correspond to the intrinsic property of inertia responsible for manifesting the gross universe? Could discoveries like these help us to learn more about the causal universe? Regardless of the actual answers to these questions, no doubt incorporating the schema of the universe detailed here into ongoing rigorous scientific inquiry *will* lead to meaningful discoveries—perhaps even to the grand unified theory of everything that has so far remained elusive.

With this, our accounting of the origins of the manifest universe is complete. This two-phase event did not spawn from nothing, but rather from the vast potential stored in the causal universe. It was triggered by a specific destabilizing event—the cumulative force of all the manifest universes that preceded it. This beginningless cycle of manifestation and unmanifestation, potentiality and actuality, the Big Bang and the Big Crunch (or Big Rip) have preceded our current manifest universe for countless iterations and will continue to do so in perpetuity.

There are still several topics we are yet to explore in full such as causality principles and the nature of infinite consciousness. In the next chapters, we will continue our systematic analysis by exploring these subjects, all the while working toward the methodology that will enable the ultimate human goal—pursuing and capturing infinity.

Chapter 20

The Way Forward

Having completed this analysis of the universe and its creation, we are now ready to move one step closer to the ultimate goal described in this book—capturing infinity. Our study will now turn to the specific methods you may adopt as you begin your pursuit of this goal.

But before we move on to a discussion of the qualifications, preparation, and practices involved in this pursuit, let's briefly summarize the path that has led us to this point; this will provide useful context for the subjects covered in the coming chapters.

How we got here

We began this discussion by identifying the rare 10 million individuals already engaged, in one way or another, in the pursuit of infinity. Whether they use high-tech instruments or inferences drawn from subtler observations, these individuals process data gathered by scientific methods to drive their attempts to answer the most far-reaching questions humans are capable of asking—the *when, where, why, how,* and *what* of the universe.

It's important to note, however, that these efforts are limited to observations of the gross universe, obtained via instruments composed of gross matter. And yet, their objective is to capture infinity. Whether these efforts involve study at the macrocosmic level of astrophysics or the microcosmic level of quantum physics, the fact remains that they deal with gross matter. Both governments and individuals have directed many resources, measured in time and money, toward these gross pursuits of the infinite.

Despite all these efforts to use gross means to study gross matter, we noted that the ultimate validation tool is actually found in the observer's intellect—a component composed of subtle matter.

Beyond that, we studied the nature of the Big Bang—a discrete event occurring some 13.8 billion years ago. While recent observations from the JWST may lead to a revision of the precise date, we can still be *relatively* sure of its timing—and we can be *positively* sure of its reality. We need no further proof than to look around ourselves to recognize the reality of the gross matter that was formed during this event!

We pushed beyond the gross Big Bang to study the subtle Big Bang, which gave rise to the subtle universe—a universe comprised of subtle matter made up of the five pure states of fundamental matter, the building blocks of thoughts. We also saw how this subtle Big Bang was the result of the destabilization of the singularity or causal universe that preceded it.

Through our analysis, we eventually reached the logical conclusion that consciousness is a non-material, eternal, and all-pervading medium in which all three universes—causal, subtle, and gross—are bathed. This medium not only forms the basis for human consciousness but also for the very cycle of universal manifestation and unmanifestation itself.

Finally, we revealed how when one eliminates all their unfulfilled desires, they are united with their true identity and with consciousness itself and will achieve a state of infinite happiness. We described this event with a chain of equated concepts: **I = You = Consciousness = All-Pervading = Infinity = Infinite Happiness.**

So, how does one actually achieve this state? We examined the relevant concepts and theories—but what *method* does one use to pursue and capture infinity? As we are about to see, this is as much a question of *how* as it is of *who.*

Who will proceed?

At this point, readers of this book will generally fall into one of a few categories. Some will dispute what we have explored up to now and will either remain skeptical or write these insights off entirely. Others will have a more favorable view of the ideas discussed here but will feel that they have gotten everything they need from this book and will likely set it aside, perhaps only returning to it now and again for reference. In both cases, such individuals are not likely to continue to build on the lessons they've learned here.

There will be a smaller group of thinkers who find this material provocative enough to try to incorporate some of these concepts into their existing scientific practice. This is an important and necessary use of these ideas and is likely to lead to significant advances in mankind's understanding of the universe.

Finally, the smallest segment of readers—a mere sliver out of the already rare 10 million—will be compelled by the content of this book to dedicate themselves seriously and completely to the pursuit of infinity. Dedication to this pursuit involves a genuine, logical faith that the total dissolution of thought will result in a state of unity with the infinite.

Ideally, all the readers of this book—and, indeed, all of the rare 10 million intellects—would pursue this objective.

However, this path requires dedication. As each individual is naturally inclined toward the pursuit of their existing desires, we must acknowledge that not everyone will take this up. Thus, it is to the fourth, smallest sliver of readers that the remaining chapters of this book are dedicated.

Once this rare set of people have realized that the ultimate purpose of being born human is to pursue the infinite, the next objective will be to prepare them for this attempt. Think of this preparation in the context of the JWST, which underwent decades of preparation and fine-tuning prior to its launch in 2021. If this gross instrument requires such extensive preparation, consider what will be involved in the preparation of one's own internal instruments—the ego, memory, intellect, and Mind—ahead of this subtle pursuit of the infinite.

While this preparation does require seriousness and dedication, it should not seem daunting or overwhelming, as there are specific, proven methods and teachings available to guide one's path. Embracing these resources will help one prepare both their subtle and gross bodies for this noble effort.

Just as one must complete numerous aptitude tests to gain the qualifications necessary to attend a prestigious university, there are numerous criteria one must meet and steps one must take to be able to actually capture infinity. Being one of the rare 10 million represents a significant qualification in itself, as it demonstrates a commitment to this pursuit. **However, whether or not one will ultimately be able to capture infinity depends on the intensity of their desire to make capturing infinity their primary—and, eventually, their *only* goal.**

Let's begin our discussion of the methods you may use to prepare yourself for this attempt with an overview of the principles and their relationship to the content of this book so far.

Science and Vedanta

Going forward, all of the methods and concepts we will discuss will have their origin in Vedanta. What is Vedanta? *Vedanta is, was, and will be a trans-scientific firmware that comprises systematic sets of eternally valid logical tools, algorithms, principles, systems, and laws.* These resources are meant to guide individuals who wish to pursue and capture infinity, offering direction as to the qualifications, preparations, and practices necessary for this attempt. In short, it is a philosophy, in the truest sense of the Greek etymology: *philosophia,* 'love of wisdom.'

So far, this book has avoided any mention of Vedanta or any related terms; **however, every concept that has been discussed up to now has derived directly from Vedanta.** This includes everything from our recent explorations of creation and consciousness, all the way back to our earlier discussion of the human composition, human goals, and the logic behind capturing infinity.

The reason why the connection with Vedanta was not introduced in earlier chapters is simple. Many thinkers, particularly in the West, are heavily conditioned to associate Vedanta (and, indeed, any similar text) with mysticism and irrationality, and to hold it as incompatible with scientific pursuits. Neutrally presenting our content allowed us to proceed from one concept to the next rationally and systematically, avoiding bias and proving the soundness of the logic behind the ideas.

As a rational system, Vedanta can be viewed as an extension of a scientific thinker's existing practice. However, once the concepts contained in Vedanta are truly understood by the same thinker, they inevitably realize that science and physics are themselves derivatives of Vedanta.

As we discussed in the introduction, where physics ends, philosophy begins. Over the centuries, however, the gap between science and philosophy has grown. As a result, the two fields are disconnected from one another, and there is little possibility of consensus. One of the intentions of this book is to narrow that seemingly ever-expanding gap. Vedanta, as a system that encompasses both science and philosophy, is an ideal means of doing this.

As we spend the next chapters discussing concepts rooted in Vedanta, it will be necessary to include some Sanskrit terminology—something this book has avoided up to now. The use of these terms is not meant to confuse the reader, and they will always be thoroughly defined. Ideally, their inclusion here will both clarify their contemporary misuse while also helping the reader along in their preparation for the pursuit of infinity.

Foundations in Vedanta

Vedanta originates in a series of ancient texts called the *Vedas,* which are divided into four groups: the *Rig, Yajur, Saama, and Atharvana.* The end of each of the four Vedas comprises a set of very important texts called the *Upanishads.*

The Upanishads deal with concepts such as human composition, universe, creation, infinity, consciousness, human goals, their inter-relationships, how to achieve the state of infinite consciousness etc. They form the basis for the concepts already presented throughout this book.

The term *Vedanta* derives from the Sanskrit words *veda* meaning 'knowledge,' and *anta,* meaning 'end,' or 'ultimate.' Thus, *Vedanta* not only refers to the fact that the Upanishads literally come at the end of the Vedas but also that they represent the ultimate knowledge—in a very real sense, the knowledge of infinity.

Despite being a profoundly secular system of thought that applies to all humans, in modern times, the Vedas have become bound to the concept of Hinduism. The concept of a religion known as 'Hinduism' is a comparatively recent one, introduced by Western colonizers in the 18th and 19th centuries. The Vedic culture preceded this development by thousands of years and the set of people who followed the Vedic culture are called Vaidikaas, later (mis)-named as Hindus. Neither in history nor in content are the Vedas explicitly religious texts. By acknowledging this history and decoupling the Vedas from religion, specifically Hinduism, the reader will be able to access the universal, secular, scientific content of these works and realize the infinite value of Vedanta.

In addition to the Upanishads, there are two other primary Vedantic texts: The *Brahma Sutras,* and the *Bhagavad Gita.* Together, these three works are known as the '*Prasthana Trayam,*' meaning the 'Canonical Triad' of Vedanta.

The first of these texts, the Upanishads, are ancient cryptic texts that deal with consciousness, infinity, creation, and the other subjects we've already discussed. Originally, they included more than a thousand Upanishads, 1,180 by some accounts, though this number cannot be verified as so many have been lost to time. In the present day, 220 Upanishads survive. Of these, only 108 Upanishads have been interpreted by Vedantic scholars. These are considered the authentic Upanishads and are highly encoded. They can only be deciphered by well-learned Vedantic scholars.

The triad's second set of texts, the Brahma Sutras—literally meaning 'formulae of the infinite'—collects the messages strewn throughout the Upanishads and restructures them into a codified, systematic presentation. Thus, if the Upanishads are strewn flower petals, the Brahma Sutras are a garland of flowers. These 555 sutras were compiled and composed by the sage Vyasa, and take the form of short, pithy statements (known as formulae) that deal with infinity, consciousness, creation, and human composition.

The third and final text is the Bhagavad Gita, the most popular Vedantic text of modern times. Also composed by the sage Vyasa, it comprises 701 verses that decode, expand on, and further systematize the contents of the Upanishads and Brahma Sutras, making them easier to understand and apply to life.

As time has progressed and human intellect has deteriorated, even the decoded Vedantic messages of the Bhagavad Gita have become difficult for the intellect to comprehend. To address this, numerous auxiliary texts have been composed throughout the years, along with commentaries and sub-commentaries provided by various Vedantic scholars.

Some of the more prominent Vedantic works include the *Vedanta Saara, Tattva Bodha, Bhaja Govindam, Atma Bodha, Vedanta Dindima, Pancha Dasi, Anubhuti Prakasika, Aparoskha Anubhuti, Drg Dishya Vivekaha, Jiva Yatra, Manisha Panchakam, Niti Shatakam, Saadhana Panchakam, Saddh Dharshanam, Uddhava Gita, Upadesha Sahasri, Vairagya Shatakam, Vaakya Vritti, Vichara Sagara, and Viveka Chudamani.*

The reader should note that it is ***absolutely NOT necessary*** to study or even know about all of these auxiliary texts. They are mentioned for readers who may be curious, but studying all of the texts is not at all required for the pursuit of infinity.

Coupled with an introductory book like this one, a thorough study of the Bhagavad Gita and a review of two or three of the Upanishads would be more than sufficient. No matter which texts one uses, the key is consistent, systematic study over a period of time, under the guidance of a competent teacher. The time required depends entirely on the strength of one's secondary desires compared with that of the primary desire for capturing infinity, and the strength of one's intellect. Regardless of that, the systematic study of these texts is one of the best ways to increase the intensity of the primary desire over time, which will help one eventually capture and become one with the infinite.

With this fundamental understanding in place, let's move on to an analysis of the prerequisites or qualifications necessary for a successful attempt at capturing infinity.

Chapter 21

Prerequisites

Just as prestigious institutions and high-ranking positions require specific qualifications for acceptance, the pursuit of infinity also needs a set of prerequisites. These prerequisites are meant to prepare one's subtle and gross body for the fulfillment of this ultimate human goal. The rare 10 million intellects already possess some degree of these qualifications, but belonging to that group is not a complete qualification in and of itself. Dedicated work will be required over a sustained period to obtain the qualifications for this pursuit.

The four-fold qualifications

Vedanta specifies a set of four primary qualifications for the pursuit of infinity, known as Saadhana Chatushtayam—*Saadhana* meaning 'qualifications' and *Chatushtayam* meaning 'four-fold.' The reader should not get distracted by these Sanskrit terms, which are only provided for context and completeness. As we move through our explanation of each qualification, all terms will be defined in clear, secular English.

First qualification: Intense burning desire for infinity (Mu-Muk-Shatvam)

The first of the four qualifications is the intense, predominant, burning desire to explore and become one with the infinite self—and, in doing so, with infinite consciousness. This intense desire, known as *Mumukshatvam,* is also associated with happiness. As we have discussed, because consciousness is happiness, the pursuit of this goal must also be characterized by happiness.

Recall that the pursuit of infinity is only one potential human goal. There are three other categories of goals as well—goals of security, comfort, and invisible wealth. As the desires associated with these goals are impermanent and changing, satisfying them only brings temporary happiness. All goals of this nature have inherent defects that lead to dissatisfaction, painful attachment, and sorrow.

Individuals who only pursue goals of these three types are bound to suffer constantly. In an attempt to alleviate this suffering, they will constantly shift their goals. This may sometimes bring temporary relief but ultimately leads to a state of intractable entanglement. Worse still, this entanglement extends beyond one lifespan and will doom the individual to innumerable lives of suffering throughout the cycle of birth and death.

Vedanta describes this state of helpless entanglement and suffering in one word—*Samsaara.* This word refers to the act of latching oneself to the impermanent aspects of life, becoming dependent upon them, and suffering the ill effects of this commitment throughout the cycle of birth and death. This suffering may manifest in the form of insecurity, worry, anxiety, fear, frustration, disappointment, hatred, jealousy, anger, lust, delusion, greed, possessiveness, selfishness, and so on. All these conditions result from the expectation of permanence from the satisfaction of impermanent goals—a deeply misguided understanding of the nature of worldly pursuits.

However, by building a very strong desire for the pursuit of infinity, one systematically frees oneself from this cycle of suffering. This is why it's so important to set the fourth human goal, the goal of capturing infinity, as one's primary goal. It helps to build momentum toward breaking free from this cycle.

A pair of examples from Vedanta will illustrate the importance of Mumukshatvam, the strong desire to gain freedom from the state of entanglement and suffering.

First, imagine that you are pushed into a deep pool of water. Your head is forcibly held underwater from above. What would be your state of mind? **Your only thought at this moment would be to somehow get your head out of the water.**

Next, suppose you are home alone, and your house catches fire. Imagine your state of mind as this conflagration rages around you. **Your only thought would be to get out of the house.** You wouldn't spend time sitting around and thinking of ways to save your toiletries, kitchen utensils, and other gadgets.

Now consider this—being born as a human is the equivalent of being thrown into that pool of water or being trapped in that burning house. The single-minded desire to get out of the water or to get out of the fire can be equated to the strong desire to pursue the infinite. The pursuit and capture of the infinite is the only way to escape from the cycle of entanglement. If one does not escape, they are destined to be born back into the same burning house. Such is the significance of Mumukshatvam, the strong desire to gain freedom from entanglement and suffering.

The thought of mustering this level of intensity may seem daunting, but it doesn't have to be. Even a weak intensity is enough to get started. With commitment, that intensity will build over time. Even committing 5 to 10 percent of one's daily thoughts toward intensifying the desire to capture infinity would represent a significant commitment. Over time, sustained commitment will progressively lead to higher degrees of intensity.

Freedom from Samsaara is known as *Moksham* or *Mukthi,* which refers to the unity of the self with infinite consciousness. At the macrocosmic scale of the universe, this infinite consciousness is known as *Brahman;* and when manifested on the microcosmic scale, in a single individual, it is known as *Atman.*

The reader should note that the term Brahman is not to be misunderstood as meaning the same thing as *Braahmiin,* which refers to the priestly caste within the Indian caste system—a system that was designed to benefit humanity, but which turned into the worst imaginable curse on the Indian community. In the next chapter, we'll briefly address the caste system and its relationship with the rest of the ideas presented in this book. For now, note only that the terms are not identical, but that any individual (not necessarily only those born to a Braahmiin family) who sincerely and systematically dedicates their life to the pursuit of infinity (Brahman) is known as a Braahmiin.

Second qualification: Discrimination (Vivekam)

The second qualification for the pursuit of infinity is discrimination. Discrimination here does not mean 'prejudice.' Instead, the term *Vivekam* means 'intellectual discrimination,' the capacity to think, analyze, reason, judge, validate, decide—in short, to use logic.

Discrimination is the capacity of the intellect to first understand and then negate all the material, impermanent aspects of the world. It's also the capacity to understand the permanent aspects of the world, including those within one's own personality—the infinite consciousness that is present within all of us.

Discrimination enables one to use the temporary, fleeting aspects of the world to fulfill the temporary needs of security, comfort, and invisible wealth—but to rely on them for *no more* than that. Through the discriminating application of one's intellect, one may create opportunities to advance one's education, derive entertainment, and build reasonable material wealth. However, it's necessary to have a strong conviction that one's current happiness is not dependent upon future accomplishments, nor is it dependent upon any other temporary aspects of the world.

When we look beyond the fleeting conditions of this world, we can discern three universal factors that appear to be permanent—time, space, and causation. However, these seemingly eternal factors are not permanent. **Anything that *is*, is not permanent. Once one unites with *real*, all-pervading permanence, one realizes that even time, space, and causation are temporary.** This realization depends upon the strength of one's intellect.

Again, one need not possess this powerful level of intellectual discrimination right at the beginning of their journey. Simply attempting to discriminate between permanent and impermanent aspects of life, and to assign the proper values to these aspects, is a start. This ability will develop over time until one can use intellectual discrimination to help support their goal of reaching the infinite consciousness.

Third qualification: Dispassion (Vai-Rag-Yam)

Dispassion or *Vairagyam,* is the third qualification for the pursuit and capture of infinity. It is a widely misunderstood concept of Vedanta. It does not mean that one has no passion. Instead, it means that, based on one's intellectual strength and inner resolve, one has a mastery over their sensory experiences and cannot be emotionally swayed by the sensory stimuli of the world.

This cannot be achieved by running away from the physical, sensory, impermanent world. As long as one inhabits a gross body, one will *always* belong to the sensory world. So, such efforts are doomed to fail. Instead, one attains dispassion when one can fixate their thoughts on the one permanent goal and is not distracted by sensory things.

To illustrate this point, consider the sandcastles built on the beach. They may be pleasurable to make and look at, but it would be absurd to place one's safety and security in the idea of living inside one. So, it is with dispassion—one may take pleasure in temporary things, but cannot fixate upon them as if they are permanent. When one lives inside a real castle, one may still enjoy sandcastles, but will not be affected when they are washed away by the sea.

Dispassion is all about placing the right value on sensory objects and emotional relationships, which can help fulfill secondary goals of security, comfort, and invisible wealth. All the while, though, one must never lose sight of the primary and permanent goal of capturing infinity.

By attaching oneself to that highest goal, one will inevitably begin to perform their actions in this world with a sense of dispassionate detachment. As this begins to occur, one will derive happiness and peace during their travels along the path and will not be affected by the failures or setbacks one faces in pursuit of the secondary goals.

This tongue twister encapsulates the point nicely, as it applies to both the relative and secondary goals, as well as the absolute, primary goal. "Attach to the attachment of the detached, to get detached from the attached."

Fourth qualification: Six-fold wealth (Shatka Sampathi)

Regardless of whether one is pursuing secondary goals or the primary one, as long as they are part of the world, they will have to take action and experience situations involving impermanent beings and objects. It is essential to be able to manage them effectively and harmoniously. This ability is the fourth qualification, known as *Shatka Sampathi* or 'Six-fold wealth' referring to inner wealth that comes from inner discipline. This inner wealth is vastly superior to material wealth.

To practice this, one must first hone their self-discipline; then, they can apply that same inner discipline to their dealings with the wider world. Managing one's inner personality by developing a strong intellect and a disciplined mind oriented around higher values is an essential prerequisite to capturing infinity. Achieving this depends on one's commitment to inculcating and living by the following six attributes of inner wealth.

- ***Equanimous mind (Samam)***

The term *Samam* means 'self-control' or 'mind control'; however, as the word 'control' implies restriction and suppression, a better definition is 'an equanimous mind.' Vedanta doesn't advocate any outright suppression, as this is guaranteed to lead to harmful consequences for the inner personality. Therefore, Samam refers to a regulated, moderated, and equanimous state of mind.

As we know, the mind is the seat of our desires and emotions, and it can generate a constant stream of new thoughts. This constant generation of flow of thought is the nature of the mind.

If one attempts to suppress the thoughts by ignoring them or refusing to acknowledge them, this will do nothing to *stop* the flow. It will only create a dangerous buildup of thoughts in the mind that will eventually lead to emotional disturbances such as stress and anxiety. Such disturbances will hold one back in their pursuit of infinity.

To cultivate a calm, composed, equanimous mind, one must develop and exercise a strong intellect, as the intellect is the component of the inner personality responsible for regulating the mind. Thus, though this aspect of inner wealth refers to an equanimous mind, it really depends on a strong intellect.

If you're wondering how to develop the strength of intellect necessary to achieve an equanimous mind, recall that at present we're only discussing the qualifications for pursuing infinity, *not* the manner of actually *acquiring* those qualifications. Later chapters will be devoted to the specific practices one should pursue in support of these efforts.

- ***Sense regulation (Dhamam)***

In the previous chapters, we discussed how the five gross sensory organs are constantly directed toward the world, and how during the waking state, the five power centers of perception translate sensory stimuli into the raw material of thought.

Each sensory stimulus experienced by the body has the potential to be translated into a thought in the mind. With all five sense pathways wide open throughout waking life, this represents a potentially torrential flow of new thoughts *just* from sensory perceptions.

Thus, Vedanta advocates *Dhamam*, the regulation of the sense perceptions. This is a preventive measure that is intended to filter out unnecessary sensory inputs. This reduces the total flow of potential thoughts in the mind and makes it easier to focus on sensory perceptions that are more relevant to one's goals.

- ***Withdrawn mind (Upa-Ramam)***

The mere fact of setting the goal of capturing infinity does not eliminate the other secondary goals from one's experience of life. Even for one who is highly devoted to the fourth goal, a few of the other three goals will always be lurking around on the periphery.

The practice of *Upa-ramam*, or mental withdrawal, helps prioritize one's dedication to the fourth goal by withdrawing one's dependency on external, worldly objects. It enables one to appropriately divide focus among various activities, dedicating sufficient time to the fourth goal of infinity while allocating a reasonable, minimum amount of time for secondary goals. In this respect, it is related to both Samam and Dhamam, in that it involves reducing, but not eliminating or repressing one's focus on second-order priorities.

With a withdrawn mind, one is focused on performing their obligatory duties, which are directly relevant to meeting set goals. Such a person will become introverted, in that the focus will be on the pursuit of the goals rather than satisfying a dependency on external objects. *Introvert* and *extrovert* do not mean the same thing in this context as they do in general usage. Here, an extrovert depends on the objects of the world for the present contentment; an introvert does not depend on such things and is instead focused on performing actions that help in getting closer to the goals.

- ***Mental resilience (Thi-Thik-Sha)***

The experience of life is varied, and at times involves physical pain or emotional distress. Whenever we suffer such pains, it's natural to immediately seek some solution to bring about their end. We may succeed in alleviating pains temporarily—but it is inevitable that, in due course, some other pain will crop up. After all, life is a series of experiences, and pain and suffering are a natural part of it.

Some may attempt to ignore these pains, only to have them reach a tipping point—at which time, all the pent-up discomfort will spill out in disproportionate anger or some other negative behavior. Others may succeed in suppressing their pain completely, only to experience depression or other chronic damage to their personality.

Everyone has a different degree of resilience when it comes to enduring physical and emotional pain; what is painful for one may mean nothing to another. This indicates that different thresholds for experiencing pain exist, and up to that threshold, pains are not experienced *as* pains.

Therefore, a more durable means of overcoming pain is to raise one's level of resilience. This means increasing the threshold for experiencing pain and reducing the degree to which one identifies with pain so that there is faster recovery. By increasing the ability to withstand pains, one is better equipped to endure them equanimously and even treat them with ease and pleasure.

However, this does not mean that one is not bothered by troubling experiences—particularly ones that involve harm to oneself or others. Distancing oneself from the experience of pain has nothing to do with allowing oneself to be taken advantage of or harmed. The same intellect that enables one to forbear pain can also help enforce appropriate boundaries and defend against encroaching forces.

- ***Intellectual faith (Shraddha)***

One's ignorance of infinity will continue up until the moment that one finally captures it. Thus, no matter how advanced one is in their journey toward capturing infinity, they will always be somewhat ignorant of the nature of the object of their pursuit. This highlights the importance of intellectual faith or *Shraddha*.

This book has covered many abstract subjects, from the logic behind capturing infinity to the nature of consciousness. The reader has often been encouraged to simply accept what is written and continue the study with an open mind—but, at the same time, not to accept what is written on blind faith. The best practice is to proceed by neither rejecting unfamiliar things nor accepting new information uncritically.

This is the nature of intellectual faith—the ability to continue one's study without intellectual arrogance, and to focus on developing one's intellect and knowledge until the point at which one achieves conviction and complete understanding.

Two quotes from renowned poets succinctly and beautifully express the nature of intellectual faith:

> "A belief in a thing which I do not know until I come to know what I believe in."
>
> – Joel Goldsmith

> "Faith is the bird that feels the light and sings when the dawn is still dark."
>
> – Tagore

Intellectual faith can help one acquire the principles associated with Vedanta more quickly, allowing one to proceed along their journey more quickly, too.

- ***Tranquil mind/concentration (Samadhanam/Dharana)***

Concentration and focus have long been components of management jargon and have been used by basically every motivation guru across the globe. Everyone agrees that concentration is important, but what does it actually mean?

Recall the definition from earlier in this book: concentration refers to the intellect directing the mind and the power centers of perception and action to continue engaging in the present activity. When concentrating, the intellect does not allow the mind to scatter thoughts into the past or the future but rather keeps them focused on sustaining the present activity until its satisfactory completion.

To concentrate, one must have clear goals with well-articulated milestones. In our case, the goal is the ultimate human goal, and the milestones can include attaining the prerequisites described in this chapter. Because this is the most significant goal, our pursuit requires much more concentration than would be necessary for secondary goals.

Objectivity can be considered a corollary of concentration. It involves a steady, continuous application of the intellect throughout all aspects of life's activities. This requires a robust intellect capable of registering and reacting to thoughts objectively.

With this, our analysis of the sixth aspect of inner wealth, along with the fourth overarching qualification, has concluded. Let's wrap up this chapter with a look ahead at the structure of the coming chapters, which will detail methods the reader may use to gradually enhance their ability to pursue infinity.

Looking forward

To effectively pursue and capture infinity, one must focus on attaining each of the above qualifications using a systematic, proven method.

Anyone reading this book is bound to already have some measure of each of the above qualities. To get to this point, one must possess a reasonably strong intellect, which will be a huge asset going forward.

However, the intellect isn't everything. Remember that the human personality consists of several layers. Just as each state of gross matter inevitably contains the others, all layers of the human personality are inevitably present in each individual. Even when one layer predominates—for example, the intellect—one's personality invariably contains greater or lesser degrees of the mental layer, the active layer, and the physical layer. All these layers in tandem with the number of desires that correspond to each of them combine to define one's overall personality.

Vedanta takes all these variables into account to provide guidance that will uplift and refine all aspects of one's personality. This guidance takes the form of systematic, logical, active procedures known as *yogas*. Different yogas are prescribed to different individuals in varying proportions and for varying durations, depending on the state of one's overall personality.

Most people will already be familiar with the term yoga, though they're likely to think of it as a practice that has more to do with physical exercise and meditation than supporting one's ability to achieve their goals. However, as we've remarked more than once, yoga and meditation have become highly commercialized, both in the East and the West—to the extent that they have lost much of their meaning for many people.

For the next several chapters, we will address the true nature of the various yogas, along with a study of the specific yogas that Vedanta prescribes in specific cases. But first, let's briefly examine the caste system (which came up earlier in this chapter), since there is some important overlap between that system and the broader lessons described in this book.

Chapter 22

The Cost of Caste: Blessing Turned Curse!

The caste system was devised as one of the greatest blessings to humanity—a perfect social justice system that would serve to help individuals satisfy all four categories of human goals, culminating in the achievement of a state of infinite consciousness

Instead, the caste system has become the greatest curse imaginable. It is in shambles and has contributed to a society that is totally at odds with its original intent.

In this chapter, we're going to address the societal caste system (*Varna Dharmam*)—both as it was intended, and as it is now—in addition to the individual lifestyle system (*Aashrama Dharmam*) that was intended to go hand-in-hand with it.

For the reader who is exclusively interested in developing the prerequisites for capturing infinity, this chapter is *not* a direct prerequisite for that goal. However, for those interested in learning the vital context of this social system and how it relates to the lessons presented throughout this book, this chapter will briefly present some important information before we move on with the rest of our discussion.)

One additional note on terminology: while we've generally avoided Sanskrit terms throughout the book thus far, this chapter is full of them on account of the subject matter. As we mentioned previously, it is not necessary to learn or retain these terms to proceed with the goal of this book. They are provided only for context and completeness.

Putting caste in context

Over the centuries, the caste system has evolved to earn the reputation of a rigid social code that sorts people into a strict hierarchy based mainly and only on the circumstances of their birth. This inflexible system has been wielded by self-interested parties (originating from both within and outside India) to sow division and to enable the perpetuation of colorism, classism, and other baseless biases.

And yet, despite being responsible for inflicting suffering based on the material conditions of this world, this system has its origins in the same texts that gave us the path toward capturing infinity.

How could it be that ideas that enable both subjugation and liberation could have the same source?

To understand that, we need to look at the caste system in its original context.

The origin and purpose of the caste system

To begin, let's return to an idea we've focused on throughout this book—the various types of human goals. Recall that there are four categories of human goals:

- **Goals of security, including food, clothing, shelter, and health (*Artham*)**

- **Goals of comfort, including material, physical comforts and luxuries (*Kaamam*)**
- **Goals of invisible wealth, or actions taken in this life to manifest rewards in the next one (*Dharmam/Punnyam*)**
- **The goal of infinity, or merging with infinite consciousness/happiness (*Moksham/Brahman*)**

While the fourth goal, Moksham, is the **primary** goal of all human beings, Vedanta also has a term that refers to these four goals **collectively**—*Purushartham* (Purusha-Artham), where *Purusha* means "human being" and *Artham,* in this context, means "meaningful purpose."

In a previous chapter, we mentioned that the means of achieving Moksham are recorded in a set of ancient texts known as the Upanishads, which deal with concepts such as infinity, creation, human composition, the nature of the universe, and the ultimate knowledge of consciousness. They, in turn, are the end sections of a larger set of ancient texts known as the Vedas. This may have led you to wonder about the importance of the initial parts of those texts.

The earlier sections of each of the Vedas are known as the *Veda Poorvam*. These sections comprise various rituals and value systems aimed at the betterment of humanity. While the Veda Poorvam includes many valuable systems and practices, one that stands out in particular is the *Varna-Ashrama* system—a system designed to help both individuals and society as a whole to pursue *Purushartham,* the collective term for the four human goals.

Why are we focusing on this system now? Because the Varna-Ashrama system contains the caste system. Specifically, it is a two-part system that consists of:

- **Varna dharmam**, or the societal caste system, which was originally intended to h**elp society in general to grow and prosper.**
- **Aashrama dharmam, or the individual lifestyle system, which focuses on the ind**ividual (rather than social) growth.

Together, the two halves of the Varna-Ashrama system were intended to uplift all of humanity, independent of religion, helping society as a whole to pursue and achieve the four human goals.

As described in the Veda Poorvam (and in the *Manu-Smriti,* an ancient legal text), the Varna-Ashrama is a perfect system, which shall become clear, after reading the remainder of this chapter. Unfortunately, the Manu-Smriti has now become a political tool in India, misused both by those who are for it and against it. And the result has been the total collapse of the caste system, with consequences that could not be further from its original intent.

If implemented diligently, the Varna-Ashrama system of the Veda Poorvam would help every individual and all of society to attain basic material security (Artham), enjoy a comfortable lifestyle (Kaamam), achieve good deeds (Dharmam / Punnyam), and eventually accomplish the ultimate objective of all humans: liberation and salvation by capturing infinity (Moksham).

To understand what it would look like if the system were applied diligently—and to understand just how far off course the caste system has gone—we'll take a look at the two halves of the Varna-Ashrama system.

Varna Dharmam–the societal caste system

Most people from the Indian subcontinent—and, indeed, many people around the world—will already be aware of the Indian caste system, which comprises four levels:

- ***Braahmanan***, the priestly caste;
- ***Kshatriyan***, the caste of leaders and warriors;
- ***Vaishyan***, the caste of merchants and business houses; and
- ***Shoodran***, the caste of manual labourers.

But many people—even many people *within* the Indian community—may be unaware of the basis for the distinctions between the castes.

A little later in this chapter, we'll get into what the system has become. First, however, we need to address the rationale behind the original caste distinctions.

According to Vedic texts, three categories separate people into different castes within society. The purpose of these distinctions is to help people understand, based on their social position, how they can best apply themselves to pursue the four human goals, and specifically to get closer to the fourth human goal—capturing infinity, Moksham or salvation (meant for all humans)

The three categories are:

- Societal classification based on mental temperament (*Guna*),
- Societal classification based on profession (*Karma*), and
- Societal classification based on birth (*Jaathi*).

We'll take a look at each of them in turn now to understand how they relate to one another.

Societal classification based on mental temperament (Guna)

This first category is intimately connected with an idea we introduced during the chapters on the Big Bang—the concept of the three intrinsic properties of the universe: **knowledge, action,** and **inertia**.

These intrinsic properties pervade the entire universe, meaning they are present within each of us, to greater or lesser extents. These properties contribute to what are known as our *personality temperaments*. Recognizing the differences between each of these three temperaments is essential to understanding the original intent of the caste system as described in the texts.

Vedanta has names for each of these three mental temperaments within an individual:

- **The knowledge property is known as the *Saatvik Guna***
- **The action property is known as the *Rajasik Guna***
- **The inertia property is known as the *Tamasik Guna***

We all are a combination of all these three temperaments in varied proportions. The type of personality one has, and the type of actions one pursues in life, will depend on the *relative proportion* of these three temperaments within them. However, in each person, one particular type of Guna or temperament predominates, and that is the basis for classifying that person into one of these four castes:

- *Guna Braahmanan*

- *Guna Kshatriyan*
- *Guna Vaishyan*
- *Guna Shoodran*

Guna Braahmanan are those, whose innate tendencies lead them to seek the fourth human goal of Moksham or the infinite. Such people love quietude, contemplation, and nature, and are constantly seeking knowledge pertaining to the Brahman. A life of renunciation appeals to them, and they naturally seek solitude without feeling lonely. Learning the Vedantic scriptures and finding fulfilment in living out the values prescribed in Vedanta define this mental temperament.

Within such people, the Saatvik Guna predominates, with a smaller amount of Rajasik Guna and a much smaller amount of Tamasik Guna. *Based purely on this mental temperament, they are classified as Guna Braahmanan.*

Guna Kshatriyan are those whose innate tendencies and mental temperament lead them to be highly dynamic, action-oriented people. Such people meticulously plan their actions and execute them systematically, efficiently, and swiftly. They are extremely unselfish, interested only in serving, protecting, and uplifting their community. Though they are not committed to the full-time study of the vedantic scriptures, they do value the lessons found in them and will willingly serve those who *are* committed to Vedantic study—namely, the Guna Braahmanan.

Within such people, pure Rajasik Guna predominates, with a smaller amount of Saatvik Guna and a much smaller amount of Tamasik Guna. *Based purely on this mental temperament, they are classified as Guna Kshatriyan.*

The next is Guna Vaishyan. As with Guna Kshatriyan, people with this Guna also have a highly dynamic, action-oriented mental temperament. They are also capable of meticulously planning and efficiently executing their actions. However, the difference is that their dynamism is infested with *selfish* interests. They only work hard to serve themselves and their immediate family or at the expense of others in a few instances. Such people are not concerned with what is good for their country or for society at large, and at times will engage unlawful means to acquire material wealth to serve their own selfish interests. They detest Vedantic injunctions and are averse to the core teachings of the scriptures.

Within such people, impure Rajasik Guna predominates, with a smaller amount of Tamasik Guna and a much smaller amount of Saatvik Guna. *Based purely on this mental temperament, they are classified as Guna Vaishyan.*

The last of these classifications is Guna Shoodran, whose innate tendency and mental temperament is toward laziness and consumption. Such people are not motivated to earn a basic livelihood and are highly inert and irresponsible. Often drawn to drugs and intoxication, they lead unclean, unhealthy lives and create a burden on their families and society in general. They depend on others to provide them with basic amenities and lack the motivation to even pursue *Artha Purushartham,* the first human goal of food, clothing, shelter, and health.

Within such people, Tamasik Guna predominates, with a smaller amount of Rajasik Guna and a minuscule amount of Saatvik Guna. *Based purely on this mental temperament, they are classified as Guna Shoodran.*

The four castes of this category can be summarized as follows:

- Satvik predominant Guna Braahmanan

- Pure Rajasik predominant Guna Kshatriyan

- Impure Rajasik predominant Guna Vaishyan
- Tamasik predominant Guna Shoodran

Remember, though, that this category is purely based on mental temperament, and is just one of three societal classifications that make up the broader caste system.

Societal classification based on profession (Karma)

The second of these categories can classify people based on the profession they choose in life.

Ideally, the profession one chooses should be based on one's predominant Guna. However, due to family influence, societal pressure, one's own choice, or other circumstances, people are not always able to enter a profession that aligns with their mental temperament.

Regardless of *how* they entered their profession, or *Karma* (which, in this context, means 'profession'), one's livelihood determines their caste in this second classification system. The four levels within this system are:

- *Karma Braahmanan*
- *Karma Kshatriyan*
- *Karma Vaishyan*
- *Karma Shoodran*

Karma Braahmanan are those who devote themselves to studying and teaching Vedic knowledge. Because the subject matter is extremely vast and subtle, those who take up this Karma must be highly dedicated and discerning. Recall our earlier references to the 'Canonical Triad' of Vedantic texts—the Upanishads, the Brahma Sutras, and the Bhagavad Gita—in addition to the numerous other auxiliary Vedantic texts we could only name because there are too many to discuss them in detail here. And those are only limited to Vedanta; the Veda Poorvam are an entire ocean of knowledge unto themselves. Because the Vedic texts are so extensive and intensive, each section and subsection requires a specialist qualified in that specific text. Imparting Vedic and Vedantic knowledge to deserving seekers is a very noble Karma or profession; one who chooses to pursue it is a Karma Braahmanan.

The next of these castes is the Karma Kshatriyan. This is the caste of those holding governmental and administrative positions, heads of defence departments, police departments, and justice systems, leaders of businesses, media outlets, and social organizations, along with the various managers and departmental leaders within this type of organization. By and large, the objective of the above professions is to govern society, uplift the people, create wealth and prosperity, and serve the greater good. These kinds of professions require action-oriented and public-minded individuals. One who chooses to pursue this Karma is known as a Karma Kshatriyan.

The next of these castes is the Karma Vaishyan. This is the caste of those engaged in private wealth creation and distribution. Unlike those of the Karma Kshatriyan, members of this caste choose to do business primarily for their own interests. One who pursues this Karma is known as a Karma Vaishyan.

The last of these castes is the Karma Shoodran. This is the caste of those who serve the other three professions. It includes those working as labourers, semi-skilled workers, routine mechanical workers, and so on. The primary feature of these professions is that they do not involve too much thought, but rather are physical in nature. One who pursues this Karma is known as a Karma Shoodran.

Societal classification based on birth (Jaathi)

The third and final of these categories is the most misunderstood one of all—the classification of individuals based on their birth or Jaathi.

Purely based on the family one is born into, a person is classified into one of the following castes:

- *Jaathi Braahmanan*
- *Jaathi Kshatriyan*
- *Jaathi Vaishyan*
- *Jaathi Shoodran*

Why do we say that this category is misunderstood? Because the caste system based on birth has come to imply that the conditions of one's birth alone define who and what one *is*. This idea could not be further from Vedantic principles and is contradicted by the existence of the other categories as well.

For example, say you are born into a Braahmanan family, take up a position in the government, and use your influence to amass personal wealth through corrupt means. What caste would you belong to?

You are a Jaathi Braahmanan, Karma Kshatriyan, and Guna Vaishyan; you belong to all three castes at the same time.

Or say you are born into a Shoodran family, take up a profession as a small business owner, and choose to donate your wealth to some charitable cause. In this case, you are Jaathi Shoodran, Karma Vaishyan, and Guna Kshatriyan, all at the same time.

When you take into account the four castes and three classifications, you can see that numerous combinations can describe a person's true nature or 'caste!'

The fatal flaw of the caste system as it has come to exist, however, is that all that variety has been obscured. It has been replaced with a strict interpretation that not only links people with the caste of their birth but also implies that their value depends *only* on the caste of their birth.

In reality, all people are equal by birth —there can be no gradation between people simply based on the conditions of one's birth or the particular caste of a family one is born into.

Similarly, all people are equal in terms of profession, too. People are not more or less valuable to society based on their job title alone.

Now, when it comes to mental temperament, there *is*, in fact, a gradation among different people. Guna Braahmanan is superior to Guna Kshatriyan; Guna Kshatriyan to Guna Vaishyan; and Guna Vaishyan to Guna Shoodran. This is because one's temperament governs one's actions and the effect they have on society in general. And, fortunately, this caste is not deterministic. With a deliberate effort, it's entirely possible for you to gradually upgrade your Guna from Tamasik to Rajasik to Saatvik.

It's also possible to redefine your caste based on your Karma by seeking a new profession, ideally one that is aligned with your Guna. And while it may be too late to do anything about your birth or Jaathi in *this* life, your efforts to upgrade your Guna to Saatvik in this life have a direct effect on your caste of birth in the next life. So, in the end, all three of these categories are fluid. We can change our caste, both in this life and in the next.

The Vedas and Manu-Smriti discuss these logical forms of societal classification, with the specific intent of creating a system for uplifting all people and all of society by educating individuals on how to recognize and upgrade their Guna.

The system is designed to help a Tamasik Guna Shoodran (who, it should be noted, could just as easily be born in a Jaathi Braahmanan family, as a Guna Shoodran) get educated, become more active in society, gradually turn Rajasik, and eventually, become a Guna Saatvik.

The tragedy is that, as a result of the actions of selfish individuals, the original intent of the Vedas and the Manu-Smriti has been undermined, leading to the collapse of the caste system. The system's goal of promoting growth, change, and social good has been lost. Now, the vast majority of the Indian community thinks that the caste system is based *only* on birth and that one's place in the world exclusively depends on the family they were born into.

This well-designed social justice system has become a system for perpetuating social injustice and has been the primary cause of numerous social clashes across India for generations.

Restoring the caste system to its original intent would be an almost unimaginable blessing, but it would also be a monumental task. And it could not be done without the restoration of the Aashrama system as well—that is, the system for individual growth, which we referred to at the beginning of this chapter.

Let's explore this second system briefly before continuing our discussion of the pursuit of Infinity.

Aashrama Dharmam–the individual lifestyle system

The Aashrama system is designed to educate individuals on how to conduct themselves throughout different stages of life, to attain a predominant Saatvik Guna.

It focuses on systematically perfecting and upgrading the lifestyle of an individual, helping them to attain the four Purushartham as they pass through four distinct stages of life.

- ***Brahmachaarya Aashramam*:** Lifestyle of a student/ beginner
- ***Gruhastaa Aashramam:*** Lifestyle of a householder/ family
- ***Vaana-Prastha Aashramam:*** Lifestyle of a hermit
- ***Sannyaasa Aashramam:*** Lifestyle of a renunciate

Many Indian households know these four stages of life by name alone, not realizing that they each refer to an entire lifestyle.

Etymologically *Shramam* means 'effort,' while *Aa-Shramam* means 'effortless' or 'ease.' This is because the four stages are devised for the individual to enjoy their progress in every stage of life, and to transition from one stage to the next in a seamless procession. Any effort exerted in one stage builds into the next stage of life, making the process smooth and easy—which we'll see, as we walk through each of those stages now.

Brahmachaarya Aashramam: Lifestyle of a student/ beginner

In this stage, an individual receives a foundational education that sets up the rest of the life stages. Vedanta advises this mandatory early education for two purposes. The primary purpose is to refine the personality and to inculcate religious and higher values of living from a young age. The secondary purpose is to learn prescribed, specific skill sets to help people earn a livelihood and become materially independent.

Based on your own innate tendencies, you are free to choose any vocational skill set that appeals to you and to take it up as a profession. There is no limit on which of the four skill sets you may pursue, the choice is independent of the caste you are born into. For example, you can be Jaathi Vaishyan, but you are free to choose skill sets to become Karma Kshatriyan or Karma Braahmanan. Equal opportunities are provided for all to choose the vocational skill sets of their innate tendencies. In other words, you choose your vocation based on your existing mental temperament.

At the same time, all students are given equal access to a personality refinement program, to upgrade their existing mental temperament from Tamasik to Rajasik to Saatvik. To get into the program, you must choose Moksham as your ultimate goal and develop a devotion to the deity of your liking (*Ishta Daivam*). Students are taught *Hatha yoga,* physical exercises, and *Praanaayamam* breathing techniques to help maintain a fit and healthy body, ensuring longevity of the gross body. (The specifics of the various yogas and other techniques will be discussed in the following chapters.)

Students are also taught the eternal values of life or *Sanatana Dharmam*. As the name suggests, these eternal values are a set of value systems applicable in all periods of time; these values were the same before the Big Bang, in the Ice Age, 2000 years ago, and they will be the same 2000 years from now. After completing this lifestyle stage, the student moves on to the next lifestyle: Gruhastaa Aashramam.

Gruhastaa Aashramam: Lifestyle of a householder/ family

The second stage is that of the householder, also known as the family stage.

After acquiring the necessary skill sets and value systems in the first stage of life, a student enters into the sacred married life. Based on their chosen professions, the householders (both the wife and husband), earn a livelihood righteously, leading a religious life. While keeping Moksham in mind as their primary goal, they go about fulfilling their needs of material wealth (Artham) and together they enjoy various comforts of life (Kaamam) within their capacity.

The householders lead an active lifestyle, defined by Karma yoga. Gruhastaa Aashramam is the ideal lifestyle to implement and practice Karma yoga. It is a lifestyle designed to exhaust all your worldly desires in a righteous way; wherein a Tamasik temperament or an impure Rajasik temperament is refined to become a pure Rajasik temperament.

This second stage is very important in life, as it is the stage where the material wealth that contributes to and sustains the other three stages is generated (since those lifestyles do not provide any avenue to earn material wealth to meet life's basic necessities).

Vaana-Prastha Aashramam: Lifestyle of a hermit

The third stage is that of the hermit. Having enjoyed the householder lifestyle and living a life defined by Karma yoga, you are ready to transition to a stage of mental withdrawal from all worldly activities. During this third stage, there is no active effort to earn material wealth and there is a drastic reduction in worldly activities. You will be preparing and conditioning your mind and intellect to be in line with your ultimate goal of MOKSHAM at this stage of your life.

This stage is marked by the practice of *Upaasana* yoga, to upgrade your pure Rajasik temperament to a Saatvik temperament, enabling you to slide effortlessly into the fourth and final stage of life.

Sannyaasa Aashramam: Lifestyle of a renunciate

The last of the four life stages is that of the renunciate. With the sustained practice of Upaasana yoga during your Vaana-Prastha lifestyle, the mind is purified and your personality becomes predominantly Saatvik. At this stage, you mentally renounce all your worldly possessions. As a renunciate (*Sannyasii*), your life is defined by *Gnyaana yoga*. In time, you will approach your revered Guru and learn the Vedantic Scriptures for a sustained period of time. Through the process of *Shravanam* (listening), *Mananam* (reflection), and *Nidhi-Dhyaa-Sanam* (contemplating), your Saatvik personality becomes Trans-Saatvik, a state of your personality even crosses over Saatvik Guna, you become devoid of all three gunas pertaining to the world, and you at last achieve your ultimate goal of Moksham.

Of course, this is only the ideal outcome; every single person who sets Moksham as their goal does not achieve it on their first attempt. But keep the following in mind: the Aashrama system begins with setting Moksham as your deeply held personal goal, even while you are pursuing the other three human goals of Artham, Kaamam, and Dharmam. These are all indirectly contributing to the pursuit of Moksham throughout your life. Thus, the Aashrama system provides the perfect avenue to pursue the ultimate goal by allowing you to pursue it deliberately and systematically throughout all stages of life.

However, this is not the *only* path toward achieving the state of infinite consciousness. It is not necessary to have begun the Aashrama system in the earliest years of your life to achieve the fourth human goal. There are techniques that you can incorporate into your life no matter what your experience has been up to now that can help you move further along this path.

As we mentioned at the end of the previous chapter, these techniques are known as yogas—systematic, logical, active procedures designed to uplift and refine all aspects of one's personality. In the next chapter, we will briefly introduce the true nature of various yogas, before moving on to a study of specific yogas that Vedanta prescribes in specific cases. All these yogas are intended to help one systematically enhance their qualifications for engaging in the pursuit of infinity.

Chapter 23

Yoga and 'Medication'

The title of this chapter does not contain a spelling error. The fact of the matter is that many people who take up yoga and *meditation* without proper guidance and context wind up falling back on *medication* to support their emotional well-being. Why? To understand, we need to take a brief look at the way yoga and meditation are pursued in modern culture.

The term *yoga* is derived from the Sanskrit root *yuj*, meaning 'to unite, join, merge, or yoke.' It refers to a set of physical and mental practices intended to help control the mind, escape suffering, and unite with the infinite consciousness. What most people today understand as 'yoga,' however, is actually just one specific type of yoga—*Hatha yoga*. This type of yoga relies on physical techniques and is considered a preparatory spiritual practice.

The physical exercises and postures associated with yoga are known as *yoga-asana*. These practices help keep the gross body fit and flexible, to extend one's longevity so that they may have greater opportunity to pursue infinity. The discipline acquired through this yoga also helps one endure the more strenuous practices associated with the other forms of yoga.

Thus, **yoga-asana primarily benefits the gross physical body and is meant to serve as a preparatory practice on the journey toward capturing infinity.**

Meditation, on the other hand, is one of the most advanced spiritual practices involved in the pursuit of infinity. Between yoga-asana and meditation, several intermediate practices are usually skipped over by modern guides and practitioners. The leap from preparatory yoga-asana and advanced meditation is the reason why many people fail to get the benefit they hope for from these practices, and eventually fall back on medication.

So, what are those intermediate practices? There are four main ones—*Karma yoga, Bhakti yoga, Upaasana yoga,* and *Gnyaana yoga*. Except for Bhakti yoga, all other yogas, involve distinct practices and affect the human personality in distinct ways. Recall the several layers of the human personality: the physical, active, mental, and intellectual layers. Each of these layers asserts itself differently in each individual and responds differently to the different yoga practices.

Through the intentional, systematic practice of each yoga, one's overall personality will gradually become *meditative*—a prerequisite for performing the advanced meditation that precedes capturing infinity.

The problem is that most guides focus exclusively on yoga-asana and meditation, completely skipping over the intermediate stages. Without knowledge of the intermediate yogas, a practitioner's intellect and personality will be unprepared for a lasting commitment to meditation. At most, they will obtain some temporary peace and calmness—though this is more likely because they've taken a break from stressful everyday life than because they've actually embraced meditative practice.

Some practitioners may even go so far as to attend special retreats, totally removing themselves from the agitations of life by traveling to remote locations to 'focus' on yoga and meditation. However, once they return to regular life, their commitment will quickly wane. After a week or so, their practice will taper off—as will the temporary benefit they got from their retreat.

Trying to resist this all-but-inevitable decline, some enthusiastic practitioners may try to *force* themselves to continue meditation, disregarding the condition of their mind and intellect and the extent of their emotional dependency on the world. A genuine commitment to meditation requires a strong intellect; a *forced* commitment ironically *circumvents* the intellect, and leads to poor results.

Without adequate preparation, attempts to meditate are doomed to fail. Attempting to remain committed to meditation by suppressing thoughts can result in the kind of rampant, unchecked flow of thought that leads to depression or other psychological distress. This is why the modern practice of yoga and meditation so often leads to medication.

True meditation is the capacity to direct one's mind toward a single thought—toward Brahman or infinity—for a sustained period, to the exclusion of all other thoughts, under the continuous observation of the intellect. This capacity can only be achieved after diligent cultivation of all layers of the personality, practicing all intermediate forms of yoga. Only then will one be prepared to plunge into the deep Vedantic meditation that can culminate in capturing infinity!

Recall that the term *yoga* derives from a root word meaning 'to unite, join, merge, or yoke.' For two things to be merged or yoked, it implies that they are initially separate entities. In this case, those (seemingly separate) entities are the microcosmic, individual self and the macrocosmic Brahman or infinite consciousness. Preparing to unite these two entities is the object of the various yogas—the procedures, systems, and actions one takes as one attempts to get closer to achieving the ultimate human goal.

As mentioned at the start of this chapter, the primary yogas of Vedanta are Hatha yoga, Karma yoga, Bhakti yoga, Upaasana yoga, and Gnyaana yoga. These yogas address all layers of the human personality; and collectively, they prepare one for Vedantic meditation.

However, in addition to these primary yogas there are numerous other practices taught by various gurus and institutions, including Ashtanga yoga, Kriya yoga, Raja yoga, Patanjali yoga, Buddhi yoga, Dhyana yoga, Mukthi yoga, Sakthi yoga, Vipassana yoga, Kundalini yoga, Zen yoga, Tantra yoga, Mantra yoga, Vinyasa yoga, Yin yoga, Prana yoga, Hot yoga, Cold yoga, Stress yoga, Power yoga, and high-intensity interval training (HIIT) yoga. It is not necessary to learn or know about every ancillary one mentioned above!

All these practices are derivatives of the five fundamental yogas, and most of them are irrelevant to the pursuit of infinity. Therefore, for now, we will only focus on the fundamental yogas, as these enable one to enhance their qualification for capturing infinity. In this chapter, we'll briefly touch on Hatha yoga. In the subsequent four chapters, we'll examine each of the remaining yogas in greater detail.

Hatha yoga is a set of physical exercises and breathing techniques focused on keeping the gross body fit and prolonging the lifespan so that one can have the longest possible opportunity to capture infinity. Vedantic texts emphasize the importance of maintaining a fit gross physical body that will help, rather than hinder, the pursuit of infinity. However, while care for the physical body is important, *excessive* or *exclusive* attention only to the physical body should not be focused on in this practice.

Some practitioners focus excessively on the fitness aspect of Hatha yoga; they spend all their time focusing on keeping the gross body fit while neglecting the fitness of the subtle body. People fall into this trap mainly on account of vanity and an incomplete knowledge of the yogas.

At most, this will result in a healthy gross body—although this is not even guaranteed, as an obsession with the gross body can lead to unhealthy behaviors, such as anorexia, which can wind up *shortening* the lifespan rather than prolonging it. In any event, such a lopsided focus on the gross body will hinder one's progress towards infinity.

This same logic also applies to seemingly responsible lifestyle choices, like becoming a vegan. While the importance of avoiding injury to other living beings is undeniable, there are even bigger issues involved in insisting on pursuing a vegan lifestyle—namely, that it can cause distress to others. Ignorance of the effect one has on others and the world around them is a sign of a weak intellect.

Vedanta considers hunger to be a periodical 'divine disease' of the gross body, known as *Vaishvanara*. In this conception, food or *Aahaaram* is a form of medicine that serves to momentarily cure this recurring disease. Thus, food should not be a primary focus in and of itself but rather should be consumed only in a sufficient amount to provide nourishment and relieve the temporary state of Vaishvanara. The purpose of doing so is to ensure that it does not interfere with one's practice of the various yogas.

On the other side of the spectrum are those who discount the importance of physical fitness. Such people shun Hatha yoga, along with other physical fitness activities, and instead focus their energy on growing their knowledge of Vedanta. For such individuals, their gross physical bodies will become a hindrance to their efforts to capture infinity; their lifestyle may even wind up shortening their lifespan—the ultimate hindrance to this effort.

One upside of the modern fascination with Hatha yoga is that there are ample resources available, both online and offline, for learning about and practicing this yoga. The reader should explore such options to build their physical yoga-aasana practice. However, they must always remain aware of the role of Hatha yoga in the big picture—specifically, that it is a *preparatory practice* that one should engage in just enough to keep their gross physical body fit. One practices Hatha yoga so that one may be able to spend sufficient time dedicated to the remaining four primary yogas. We will now move on to discuss those yogas individually in the subsequent four chapters.

Chapter 24

Karma yoga

Along with yoga and meditation, karma is another concept that has become massively commercialized in the modern world, to the extent that its true meaning (not to mention its significance in the context of capturing infinity) has been obscured.

Karma can be understood, in this context, to mean 'proper/right action.' Thus, Karma yoga is essentially the yoga of action.

Yogas are sets of procedures, systems, and actions that one performs in their journey toward the goal of uniting with Brahman, the infinite consciousness. Karma yoga specifically does this by strengthening the first three prerequisites we discussed earlier, namely:

- **Intense burning desire for infinity (Mu-Muk-Shatvam)**
- **Discrimination (Vivekam)**
- **Dispassion (Vai-Rag-Yam)**

The latter part of this chapter will be dedicated to a discussion of how one may strengthen these prerequisites through the practice of Karma yoga.

The fourth prerequisite is the **six-fold wealth (Shatka Sampathi)** and this is developed by performing Upaasana yoga, which will be covered in a later chapter.

The interdependence of these yogas highlights the importance of practicing them deliberately and in tandem (per each individual's personality), rather than skipping directly to the end. Practicing these yogas will prepare one to ultimately move on to the practice of Gnyaana yoga, the yoga of the intellect. This final yoga represents the path of knowledge and consists of a detailed, systematic study, reflection, and contemplation of the self, Brahman, and infinite consciousness. It is only through the sustained practice of Gnyaana yoga that one can succeed in gaining *Gnyaanam* (the wisdom of Brahman) and merging with the infinite consciousness.

This orderly method has been prescribed by Vedic schools since ancient times. The reader will be interested to know that much of the content of the first half of this book has been adapted directly from components of Gnyaana yoga. Why deviate from the traditional order? There were two primary reasons.

First, this was done to appeal to objective, scientific thinkers by demonstrating the commonalities between Vedanta and their existing areas of study. Vedic texts have been unfairly associated *only* with religion and thus would generally be written off by many in the scientific community. By demonstrating the alignment in both subject matter and methodology between Vedanta, physics, and philosophy, this book hopes to overcome that bias and appeal to readers belonging to the rare 10 million intellects of the world.

Second, this inversion was done to demonstrate that they are already significantly qualified to pursue infinity, even if they've never given a thought to Vedanta before in their lives. Readers who have made it this far are clearly in possession of above-average intellects. Not only that, they are also now acquainted with many of the concepts addressed in Vedanta. Instead of keeping the lessons associated with Gnyaana yoga until the end, by placing them front and center this book hopes to prove to the reader that they not only *can* pursue this ultimate goal but that they are already well equipped to do so.

With that said, let's begin our study of the second primary yoga, Karma yoga. As the reader starts to develop a sense of the interrelation of all five yogas, along with the nature of their own personality, these lessons will reveal the specific yogic practice that corresponds to their personality and that will enable them to pursue infinity most effectively.

The logic and practice of Karma yoga

The primary objective of the five yogas is capturing infinity—otherwise known as Moksham or liberation. However, it is also important to note that the practice of these yogas will help one achieve their secondary goals in life as well. In fact, some people may even use them *exclusively* for this purpose, and not for the pursuit of infinity.

This is part of the nature of Vedanta. It is not necessary to set the highest of human goals as one's primary goal to practice these yogas. Vedanta is all-inclusive and is oriented around the rational fulfillment of all human goals.

Let's recall the three categories of secondary human goals. They are:

- **Goals of security, including food, clothing, shelter, and health (Artham)**
- **Goals of comfort, including material, physical comforts and luxuries (Kaamam)**

- **Goals of invisible wealth, or actions taken in this life to manifest rewards in the next one (Dharmam/Punnyam)**

And, of course, the fourth human goal is the primary goal of all human beings—attaining liberation through merging with infinite consciousness/happiness (Moksham). Vedanta refers to these four goals collectively as *Purushartham* (Purusha-Artham), where *Purusha* means 'human being' and *Artham,* in this context, means 'meaningful purpose.'

Throughout life, one inevitably pursues each of these goals in some measure, even without having any knowledge of this framework. In fact, it's even possible to pursue the fourth goal of capturing infinity without any technical knowledge of it!

Of course, setting the fourth goal as one's primary objective doesn't automatically dispense with the remaining three goals. It's simply impossible to eliminate goals overnight, even if doing so has become one's main objective in life. The shift toward the pursuit of the fourth goal must be systematic and gradual. Recall, from our discussion of the logic behind capturing infinity, that the fourth goal is not even an objective 'goal' in the same sense as the other three categories; rather, it is a subjective 'becoming' that consists of *eliminating* unfulfilled goals. This is a paradigm shift in the way one relates to desire—a radical realignment of one's pattern of thought.

An assessment of desire

To begin this shift, start by conducting an assessment of your desires. The goal of this self-assessment is to learn more about the nature of your personality, as your various desires each have their origin in one or another aspect of your personality. Although your intellect likely predominates, it's important to examine all sides of your personality. The pursuit of the fourth human goal involves eliminating *all* unfulfilled desires, regardless of where they originate.

If this task seems daunting, don't worry. If you wanted to lose 10 pounds of body weight, you wouldn't expect to do so overnight. You need to gradually modify your diet and your exercise regimen, and only after some time will you succeed in your goal. The pursuit of infinity is exactly the same, though the process takes even longer—because you are not only modifying your gross body but your subtle body as well. Thus, you should embrace the commitment and not worry about the amount of time involved.

During your assessment, examine the nature of your current desires. Group them based on their connection to one or another of the layers of your personality—the intellectual, mental, active, and physical layers.

As you do this, you will start noticing certain specific desires that assert themselves above others. No matter what you have in mind, these desires always recur and make themselves known. This indicates that these few desires predominate within you and are the most representative of your innate tendencies.

Vedanta refers to these innate tendencies as *Swa-Dharma*, meaning 'one's own nature.' Ideally, your profession will be in line with your Swa-Dharma and you will have chosen it early enough in life to make a substantial commitment to it. If, during your self-assessment, you find that your recurring desires are *not* aligned with your Swa-Dharma, search for a way to make some course corrections to bring your livelihood more in harmony with your innate tendency. However, avoid making any changes that would drastically affect your financial, family, or any other significant personal situations!

Why is it so important to ensure this alignment? It's because, for the reasons stated above, even once capturing infinity has become your primary goal in life, you will inevitably retain many secondary goals—and, despite your efforts to eliminate them, you will inevitably retain many of them for a long time. This is why it's best to focus on goals that correspond with your self-development and to try to dispense with those that are less consequential.

Set a realistic, unselfish goal based on the top desire or desires that you identified during your assessment. Ideally, the goal(s) will be in line with your existing profession, whether it be in the sciences, philosophy, or any other intellectual field that has the potential to benefit mankind at large.

To be clear, this concept of an 'unselfish goal' need not correspond to a *single* goal and it need not necessarily encompass the benefit of all mankind either. For example, you may have one unselfish goal associated with contributing to some scientific field; another unselfish goal associated with helping your sailing club succeed, and a third unselfish goal of pursuing your study of Vedanta. If your unselfish goals correspond with your innate tendencies, you may pursue several of them simultaneously within the available resources.

Setting unselfish goals is an important step in asserting the role of the intellect in managing your desires. The intellect is like the adult in the room, while your whimsical, childlike mind may wish to pursue this or that desire at any given time, you must assert your intellect to bring those whims in line so that you can focus more exclusively on higher-order goals. As we continue our study, we will further uncover the significance of the intellect concerning setting and fulfilling goals.

Three mental temperaments and actions

During our discussion of the caste system in the previous chapter, we introduced the concept of the three mental temperaments (Gunas)—Saatvik, Rajasik, and Tamasik—which correspond to the three intrinsic properties of the universe—knowledge, action, and inertia.

While all people have some amount of each of the three temperaments, their overall personality depends on which of the three properties predominates within an individual.

- **The knowledge property is known as the Saatvik Guna**
- **The action property is known as the Rajasik Guna**
- **The inertia property is known as the Tamasik Guna**

The type of actions you pursue depends on the predominance of each of these three temperaments within your personality. Understanding the predominance of each one is an important part of your self-assessment, as it will help you to identify patterns of behavior and focus on areas where you may wish to modify your conduct.

Vedanta classifies actions into three categories, with each category having a specific bearing on one's progress toward capturing infinity. As noted, each of these categories is also associated with a specific personality temperament. The categories are as follows.

Unselfish actions (Utthama Karma)

Unselfish actions are associated with the Saatvik temperament, which derives from the intrinsic knowledge property of the universe. These actions confer the least benefit to oneself while bringing the maximum benefit to the welfare of society. As such, these are considered the superior type of action.

Utthama Karma is performed by a predominantly Saatvik person. Such people are externally calm, but dynamic and composed inside. The knowledge component predominates within them, and they understand that consciousness is the enlivening principle of all beings. Most of their decisions are guided by the clarity of the intellect. They experience few to no dilemmas, acting instead on the power of the intellect rather than petty likes and dislikes. They maintain mental equanimity, are objective and relatively detached, and are neither affected by failures nor excited by successes. They pursue their goals systematically, without worries or anxieties, and derive happiness from the pursuit of unselfish goals.

Selfish actions (Madhyama Karma)

Selfish actions are associated with the Rajasik temperament and consist of actions taken to pursue one's interests and those of one's family, friends, and community. While these actions are not vicious and do not harm others or violate any laws, they are mostly motivated by materialistic benefits and do nothing to advance one's pursuit of infinity. As such, these are considered merely mediocre actions.

Madhyama Karma is performed by a predominantly Rajasik person. Such people are highly dynamic and their minds are infested with turbulent flows of thought and multitudes of desires. They lack the knowledge of consciousness as the universe's unifying factor and may even believe that God, an external entity, is responsible for bestowing material benefits. The action component predominates within them and they are driven by their excessive desires. They crave the temporary happiness that comes from achieving a goal and become greedy for that feeling. However, they do not derive happiness from *pursuing* their goals and may give them up entirely if they seem too onerous. Most of their actions are propelled by the mind and feed the ego, with the intellect functioning as a mere slave that may even develop a scheming nature to support this person's self-interested goals. They are thoroughly attached to the material world, suffer from constant worries, anxieties, and fears, and their behaviors may become rigid with self-imposed duties and responsibilities. Their selfishness may even lead them, at times, to unintentionally perform unrighteous actions.

Unrighteous actions (Adhama Karma)

Unrighteous actions are associated with the Tamasik temperament and consist of unrighteous attempts to fulfill one's selfish interests. Such people are rigidly convinced that their unrighteous actions are, in fact, righteous and will willingly break laws, and may even harm their own family, friends, and community, to gratify their materialistic, sensual greed. These actions bring about one's spiritual downfall, even to the extent that they may take the form of an animal or plant in the next life. As such, these are considered to be deeply inferior actions.

Adhama Karma is performed by a predominantly Tamasik person. Such people may be calm outside, but they are deeply disordered within. The inertia component predominates in them. They have no purpose in life and their intellect plays no role in their behavior. They are happy being lazy, indolent, and lethargic. They are mostly inactive and do not take on any kind of duty or responsibility, to the extent that they may become nearly inert objects. They are burdens to their family, friends, and community, and often hold extreme, rigid, fanatical religious beliefs.

In one's pursuit of infinity, it is imperative that one totally avoids Adhama Karma, reduce Madhyama Karma as much as possible, and attempt to direct the maximum amount of energy available into performing Utthama Karma i.e. proper/ right action, which is the essence of Karma yoga.

By using the above insights to guide one's self-assessment of their personality, one will be able to reorient one's behavior around actions that will help get one closer to the goal of capturing Infinity. In doing so, one would have undertaken Karma yoga, the yoga of action.

Committing to unselfish actions

Every person is a combination of the Saatvik, Rajasik, and Tamasik temperaments. However, with the right perspective, it *is* possible to reduce the share of the inferior temperaments of one's personality. This occurs when setting goals consistent with one's Swa-Dharma, which inherently necessitates unselfish actions or Utthama Karma for one's fulfillment. By prioritizing unselfish goals that are in line with one's innate tendencies and that involve unselfish actions, one ensures that one spends the maximum amount of time performing actions that bring one closer to their set goals.

One will simply not have the time to focus on other inconsequential, inferior desires any longer. Like seeds that are not watered and nurtured, over time these inconsequential desires will lose their potential to grow and will instead wither away. As this becomes an ingrained habit, such desires will arise less frequently, as the inferior temperaments of one's personality will have diminished in proportion to Saatvik—the temperament associated with knowledge and intellect.

As the intellectual aspect of your personality predominates, your actions will be less driven by inferior desires and goals. Thus, **by pursuing goals aligned with one's Swa-Dharma, one will increase their sense of dispassion and detachment or Vairagyam—one of the major prerequisites to capturing infinity.**

This dynamic will gain momentum the longer one practices it and in time will eventually diminish the proportion of inferior desires. Thanks to the increasing strength of one's intellect, one will become better able to discern between superior and inferior aspects. **In this way, one's sense of discrimination or Vivekam will increase as well—another major prerequisite to the practice of Gnyaana yoga, which ultimately leads to capturing infinity.**

Note that these prerequisites are being strengthened *without* even explicitly engaging in the pursuit of infinity; that is, even while still focused on major secondary goals, *one is nonetheless advancing on their path toward capturing infinity.*

To understand how this works, recall our formula for happiness:

$$\% \textit{ of Happiness} = \frac{\textit{Total number of desires fulfilled}}{\textit{Total number of desires entertained}} \times 100$$

By performing Karma yoga, one uses the power of the intellect to both increase the numerator (by fulfilling goals associated with one's Swa-Dharma) and decrease the denominator (by not entertaining or eliminating inferior goals). This is how the sustained practice of Karma yoga directly leads to an increase in one's happiness, along with an increase in the refinement of one's personality. (When observed appropriately, the tandem practice of Upaasana yoga can further enhance this refinement. We will discuss this practice in a later chapter).

Ultimately, by continuing this same practice with Brahman or infinity as one's goal, one will eliminate egocentric desires and the craving for the feelings associated with the results of one's actions. Instead, one receives the results of their actions with the attitude that they are divine gifts, rather than gratifying, material things.

Consider the following: When you fire an arrow at a target, as soon as you release the arrow you no longer have control over it. Similarly, as soon as you perform any action (karma), it becomes an aspect of the universe, and you no longer have any control over it. At that point, the action is simply processed by the universal, eternal laws, called *Eeshwaraa.*

Eeshwaraa can be understood at this moment as the combination of the entire observable universe in the medium of infinite consciousness. Eeshwaraa returns the processed action to you in the form of the effects or 'fruits' of your action, also known as *Karma Phalam.*

In this manner, **Karma yoga consists of the practice of dedicating your right, proper actions (Arpanam) to the eternal laws (Eeshwaraa)**—also known as *Eeshwara Arpanam*—**and receiving the effects of those actions (whatever they may be) with the attitude that they are divine gifts (Prasaadam) from Eeshwaraa**—also known as *Eeshwara Prasaadam.*

The longer you practice this yoga, the more your mind will be purified and devoid of all unwanted, inferior desires. It will have become a tool that works in tandem with the intellect in support of achieving the Purushartham or human goals.

As your personality is refined with this attitude of Eeshwara Arpanam and Eeshwara Prasaadam, the burning desire for infinity (Mumukshatvam) will become stronger. Thus, Karma yoga directly improves each of the first three prerequisites, preparing one's personality for the practice of Gnyaana yoga.

The five-fold sacrificial actions

In the ancient Vedic period, Vedanta prescribed a series of sacrificial actions, the *Pancha Maha Yagnya* or five-fold sacrificial actions as ritualistic Utthama Karma. While they may not have precisely the same relevance or applicability in modern times, they can be adapted in support of one's practice of Utthama Karma. These five actions are as follows.

Appeal to the infinity (Daiva Yagnya)

Daiva here refers to Brahman, the infinite consciousness. Despite having a personality in which the intellect predominates, you recognize the magnanimity of the infinite and bow down in reverence to it. You do not entertain the intellectual arrogance that you will 'understand' or become 'equal' to the infinite consciousness.

With heartfelt sincerity you appeal to Brahman for the welfare of all living beings and all inert objects throughout the entire extent of the manifest universe—that they may be happy, prosperous, and free from disease and suffering. This appeal is made not with gross material offerings, but rather with the subtle matter of thought.

Homage to ancestors (Pitru Yagnya)

Recognize that your parents contributed to your birth, and that *their* parents contributed to *their* births, and so on. Express gratitude for all ancestors, both those who are alive and those who have already died. For those who are alive, commit to treating them with reverence and care, and do not send them off to senior care facilities; for those who have died, perform the Vedic rituals of *Tarpanam* and *Shraddam* in honor of their contributions.

Devotion to Sages (Rishi Yagnya)

The identities of the original authors of the Vedic texts are unknown. These sages were extremely selfless. They possessed absolutely no ego and did not bother to leave their names behind. They never sought our endorsement, acknowledgement, or recognition.

Express deep devotion, respect, and reverence for all Vedantic masters of the past and present, both known and unknown—and, as these texts have been handed down for countless generations, recognize that we must preserve and propagate this timeless body of wisdom, to make it available to all humanity and ensure that it will be passed on to future generations.

Social service (Manushya Yagnya)

Without losing sight of the fourth goal of capturing infinity, participate in activities that will benefit and uplift your fellow human beings to the greatest extent possible.

In the ancient Vedic period this may have been a simpler task, as the need for broader social service was less extensive. Today, however, the disparity between the rich and the poor has widened to an almost unthinkable degree. The richest 1 percent of the population owns more than 43 percent of all wealth, while the poorest 50 percent of the population owns just 1 percent.

If society permits this disparity to continue to grow, it will inevitably lead to global unrest and civilizational collapse; through their rampant greed, the wealthy will have brought about their own downfall, along with that of everyone else.

With such a vast gulf between the rich and poor, social services *alone* are unlikely to bridge the gap. Governments and individuals should direct their resources to implement large-scale social awareness efforts, in addition to social services, since it is easier and less costly to implement at scale.

These efforts would consist of educating humanity on the nature of human goals and human composition, and on the means of securing first-order goals of security—that is, of food, clothing, shelter, and health. With a society-wide understanding of these goals, the gap between the rich and poor would be greatly diminished. This illustrates the importance of sharing the information contained in this book. As such, the reader is encouraged to share this work with others. Even sharing it with one other person will make a difference.

Protect the environment (Bhuutha Yagnya)

Recognize that you are a unit of the total global environment and that, as such, you must do your part to protect it. Extend your circle of identification beyond human beings, so that it encompasses all animals, plants, and the natural environment itself.

The most important thing we can do in this era is focus on reducing greenhouse gas emissions. Try to do so as much as possible in your own life, and support the efforts of governments and businesses to do the same for the sake of protecting the planet. Consume as little of nature as possible—only enough to support your effort at practicing Utthama Karma.

With this discussion of the five-fold sacrificial actions, our analysis of Karma yoga is complete. In the next chapter, we will discuss Bhakti yoga, an essential ingredient of every other yoga that you will practice.

Chapter 25

Bhakti Yoga

Having dealt with Karma yoga, let us now discuss Bhakti yoga, an essential yoga which doesn't exist!

How can it not exist? And, if it doesn't exist, why are we spending a chapter discussing it?

This sweeping statement will become clear as we move forward in this chapter, and especially in the subsequent chapters as well.

The logic and practice of Bhakti yoga

Bhakti can be translated as 'devotion toward God' or, in this specific context, 'devotion toward infinite Brahman.'

We can think of 'devotion' as similar to the emotion of love: the steady flow of thought that emanates from your mind toward an object or being, and an identification or feeling of oneness toward that object or being. **Devotion is nothing but the same love, only directed toward a higher principle—the infinite Brahman or God.**

Devotion can also be directed toward other higher principles, such as your favorite deity, your teacher (guru), your nation, your parents, or any person whom you revere and consider sacred.

A further analysis of the emotion of love will help illuminate the concept of devotion (*Bhakti*)—which, in turn, will help explain the meaning of Bhakti yoga.

Gradations of love

Whenever you set a goal in life and work diligently toward achieving it, love is involved.

You set the goal in the first place because you would love to attain the result. You also love the methods, means, and actions involved in working toward the goal, since they help bring you closer to the thing you love. And, most important of all, you set your goals and work to achieve them because you love yourself, and your self-love drives you to pursue and attain the things you love!

However, even though love is involved in setting, working toward, and attaining your goals, there is a gradation of the intensity of love for each of the things named above.

Namely, the love for pursuing your goal is less intense than the love for the goal itself. The love for the goal is less intense than the love for yourself in the first place.

This is because the love you have for the means of achieving your goal actually has more to do with the goal than with the means in and of themselves. Once you accomplish your goal, your love for the means will quickly be forgotten.

Similarly, the love you have for your goal actually has more to do with self-love than with the goal in and of itself, since the purpose of the goal is to bring peace and happiness. Since material happiness is temporary, as soon as you achieve one goal your love for it will fade and you will set a new goal and turn your love toward it.

Consider this example, which you may be able to relate to in your own life. Think back to your first relationship and the love you felt for the actions you took in meeting, pursuing, and bonding with your partner. Think of the extent of the love you felt for that person and the extent of the love you felt for yourself in the process. Then, recall the end of that relationship—how your goal likely shifted to finding a new, better partner… and so on.

In all your pursuits, there is a gradation in the intensity of the love you feel for the different aspects of the pursuit: a dull love for the means (*Mandha Premam*); a mediocre love for the goals (*Madhyama Premam*); and a superior love for yourself (*Utthama Premam*).

Love, devotion, and God

We can use this gradation of love to help understand gradations in the intensity of one's devotion to God. This intensity depends on your understanding of God—that is, the way you view God relative to yourself and the infinite.

The vast majority of humans who worship God tend to look upon God as a means for attaining material and emotional accomplishments for themselves and their families. This type of devotion is known as selfish or dull devotion (*Mandha Bhakti*).

A much smaller share of humans look upon God not as a means for achieving personal ends, but rather as an end in itself, i.e., they view reaching God as their ultimate goal in life, with the understanding that doing so will give them total peace, happiness, and salvation. However, even though this devotion is more noble than the selfish devotion of the majority, Vedanta considers this kind of devotion as mediocre devotion (*Madhyama Bhakti*).

Only a tiny sliver of humans have the most intense form of devotion to God—ultimate devotion (*Utthama Bhakti*). **If you have this type of devotion, you view God neither as the means for achieving material ends nor as an end in itself. Instead, you understand that God, the Infinite Brahman, is no different from you.** This can be likened to the self-love we discussed earlier. When you believe that you and God, the infinite Brahman, are at the same absolute level, you have attained Utthama Bhakti—the most intense form of devotion.

Bhakti yoga which does not exist

In the previous chapter, we covered Karma yoga, a discipline that is essential to achieving human goals (Purushartham)—including, of course, the fourth human goal of capturing infinity. In the following two chapters, we will address Upaasana yoga and Gnyaana yoga, which are also essential disciplines for achieving those human goals. And Bhakti yoga—the yoga of devotion toward the infinite Brahman—is also essential to achieving Moksham or merging with the infinite consciousness.

With that in mind, the reader will likely be eager to know what methods and techniques are prescribed by Bhakti yoga, so that they can start practicing it on their own.

The truth is **Bhakti yoga is not a distinct practice unto itself**. There is no unique set of methods or techniques that together comprise this yoga. Instead, it is the combination of **Karma yoga, Upaasana yoga, and Gnyaana yoga.** It is not a separate yoga, but rather in the common atmosphere of devotion to the infinite Brahman and consciousness in which all three of the other yogas are performed.

Those familiar with the Bhagavad Gita may argue that since it contains a dedicated chapter called Bhakti yoga, this yoga really *does* exist as a distinct practice. However, if you take the time to analyze the entire chapter, it becomes clear that Lord Krishna discusses Bhakti yoga as the *combination* of Karma yoga, Upaasana yoga and Gnyaana yoga.

In Karma yoga, it is with devotion that you dedicate all your actions to God (*Eeshwara Arpana Bhaavanam*) and accept the outcome of your actions as a gift from God (*Eeshwara Prasaadha Bhaavanam*).

Similarly, our upcoming discussions of Upaasana yoga and Gnyaana yoga will reveal how these yogas are also performed in an atmosphere of devotion (Bhakti).

Therefore, in Hatha yoga, Karma yoga, Upaasana yoga, and Gnyaana yoga, no stage is devoid of devotion (Bhakti). Thus, you may call all of these yogas put together Bhakti yoga!

Bhakti yoga in practice

All the same, even if there is no *explicit* practice of Bhakti yoga, there must be some quality that can help us to clearly recognize the difference between the gradations of devotion. Indeed, we can analyze one's degree of devotion to God in terms of their understanding of Him as the object of their devotion.

Vedanta provides three definitions of God, which depend on one's mental temperament and intellectual caliber.

If you are just starting on the spiritual path, you may understand God as the creator of the world. It is natural to recognize that, behind everything, there is a creator. Since the universe is the greatest thing imaginable, then its creator must be the greatest creator imaginable. In this concept, you visualize God as a person (or some other distinct entity), and you are devoted to that specific form. **Vedanta describes this first level of understanding and devotion as *Eka Roopa Eeshwara Bhakti* 'devotion to God with a single form.'**

As you progress spiritually and your thinking becomes clearer, Vedanta presents you with the next level of understanding of God. You will ask yourself, if God created the universe, then where did God obtain the material to do so? Even space and time are a part of it and could not have existed before it. So where did it all come from? The logical conclusion—indeed, the only possibility—is that all matter within the universe must be an aspect of God. The entire universe is nothing but God; everything and every being that exists throughout the universe is God. At this level, you have left behind the conception that God has a single form and are now devoted to this God that has many forms instead. **Vedanta describes this second level of understanding and devotion as *Aneka Roopa Eeshwara Bhakti* or *Vishwa Roopa Eeshwara Bhakti*—'devotion to God with many forms' or 'devotion to God as the Universe itself.'**

As you progress further in your spiritual pursuit and your intellect grows ever stronger, you will start questioning even this advanced understanding. You will ask yourself if God is the universe, why do we see so many disparities and so much suffering on Earth (and likely on other habitable planets, too)? Why is there injustice and pain? Why floods, earthquakes, and tsunamis? Why crime, wars, and violence? How can those things also be God?

When you ask these questions, Vedanta acknowledges your intellectual growth and presents the answer: that God is *not* the universe, but rather God only *appears* as the universe with its multitude of names and forms. This answer unlocks the understanding that God is the very substratum of the universe—an enlivening, pure consciousness, the formless principle behind the various names and forms of the entire universe. **Vedanta**

describes this third level of understanding and devotion as *Aroopa Eeshwara Bhakti*—'devotion to a formless God.' However, once your spiritual pursuit has evolved to this level, you will also show no hesitation to worship God with form, and also to worship the universe as God. Aroopa Bhakti *includes* Eka Roopa Bhakti and Aneka Roopa Bhakti!

These three levels describe the continually evolving level of devotion that you will experience as you practice Karma yoga, Upaasana yoga, and Gnyaana yoga—which, together, comprise Bhakti yoga.

In the next two chapters, we will address the practice of Upaasana yoga and Gnyaana yoga. As you read them, that initial sweeping statement that there is no such thing as Bhakti yoga will come into even clearer focus.

Chapter 26

Upaasana Yoga

In the chapter on Karma yoga, we highlighted the interdependence of the various yogas to emphasize the importance of practicing them deliberately and in tandem, rather than skipping directly to the end.

This becomes particularly clear when you recall the four-fold qualifications. While Karma yoga can help strengthen the first three of these prerequisites (dispassion, discrimination, and the intense burning desire for infinity), Upaasana yoga focuses on the fourth prerequisite, the six-fold wealth (Shatka Sampathi). This prerequisite comprises six parts:

- Equanimous mind
- Sense regulation
- Withdrawn mind
- Mental resilience
- Intellectual faith
- Concentration

The practice of Upaasana yoga can help you to develop these six qualities, making your personality (and especially your mind) fit enough to pursue and attain all four human goals (Purushartham)—including, of course, the primary human goal of merging with infinity (Brahman).

Mental fitness is essential to pursue that fourth goal. Before embarking on the path toward infinity, it is almost guaranteed that your mind—the seat of your likes, dislikes, and multifaceted desires—was in a state of constant flux. This state of perpetual change results in no end of stress, mental agitation, and other psychological burdens. Unfortunately, most people are simply conditioned to accept that this is the only way and it is normal to go through stress in life.

You can think of Upaasana yoga as a kind of *deconditioning* exercise, designed to refine and integrate your entire personality, enabling you to pursue your goals with increased efficiency and without any distractions.

Prerequisites for the practice of Upaasana yoga

Any human effort (including any yoga) is only possible in the waking state, when the gross body and the subtle body are both awake and fully functional. Thus, for your efforts (including your yogic practice) to be successful, it's necessary to refine them.

While Upaasana yoga is primarily focused on refining the subtle body, it does emphasize the fitness of the gross body as well. After all, the gross body is an extension of the subtle body, so it must be fit and healthy.

This is not the place to discuss specific physical fitness regimens, since the reader's intellect is capable of determining which practices (such as Hatha yoga and *Praanaayama*) will contribute to the refinement of the gross body. However, there is one particularly important step that we *do* need to mention prior to discussing Upaasana yoga and this relates to preparing the subtle body.

As mentioned at the start of this chapter, Upaasana yoga is primarily concerned with cultivating the mind. Since one of the predominant manifestations of one's mind is one's speech, it is therefore important to refine (or decondition) one's speech prior to practicing Upaasana yoga. This effort is known as verbal deconditioning or austerity of speech (*Vaak-Tapas*).

In short, when you practice austerity of speech, you seek to avoid causing harm with your words and instead spread goodwill and benefit others.

A verse from the great Tamil text *Thiru-Kural* expresses the importance of this idea succinctly, *Theeyinaar Sutta Punn Ullaarumm; Aaraadhe Naaviaar Sutta Vadoo* meaning "Even if a wound caused by fire is a scar on the body, the wound inside will heal in time. But the wound caused by cruel words in the heart remains a wound and will never heal."

Verbal abuse is a form of mental cruelty and must be avoided as much as possible for your own mind to be at peace. From time to time you will almost certainly find yourself in circumstances in which you might have to use strong words to drive home a point. However, it is up to you to cultivate your intellect so that you can develop methods for handling such situations appropriately without using abusive, harmful words.

Simply avoiding harmful speech is not enough . It is also important for you to make an effort to cultivate speech that is pleasant, polite, harmonious, and interesting to others.

Finally, remember that you are ultimately pursuing Brahman, which can be understood as the Absolute Truth. You can only reach it by leading a life of Relative Truth—meaning that you must do your utmost to avoid speaking lies.

As your mind becomes more subtle, you will become increasingly sensitive to lies (and their consequences); anything you achieve by lying will cause you mental agitation, while anything you achieve *without* lying will bring you happiness. Of course, there may be occasions when telling a lie might advance the interest of a noble cause that benefits others. These situations are permitted as exceptions to the rule, so long as you perform compensatory deeds (*Praaya-Chittam*) to neutralize its effects.

By cultivating your speech to be consistent with these guidelines, you will initiate an important deconditioning step that will help establish your practice of Upaasana yoga.

Before getting into the logic and practice of this yoga, let us quickly reiterate a point that has been made previously but bears repeating. In this and other chapters, various methods, instructions, and terms have been used without going into much detail about what they mean or how one can observe them. That is because the primary goal of this book is to acquaint you with these ideas and how they fit together, rather than going into expansive, elaborate discussions and explanations of every nuance.

For those interested in delving into the details of these topics, the final chapter will direct you to a point of contact e-mail, wherein you can request access to additional references, sources, and recommendations for pursuing Vedanta in greater depth.

The logic and practice of Upaasana yoga

Refinement of the mind begins with the low-hanging fruits of maintaining a fit physical body and observing verbal discipline since both of these things are nothing but gross extensions of the mind. Various Vedantic texts refer to physical and verbal discipline as being prerequisites for mental development, which is the purpose of Upaasana yoga.

Of course, refining the mind is not a goal in and of itself. It serves a greater goal—the fourth human goal, capturing infinity. Recall our earlier definition of meditation:

True meditation is the capacity to direct one's mind toward a single thought—toward Brahman or infinity—for a sustained period, to the exclusion of all other thoughts, under the continuous observation of the intellect.

This specifically defines Vedantic meditation, which will be discussed in detail at the end of the next chapter on Gnyaana yoga.

Upaasana yoga is concerned with preparing the mind to perform this advanced form of meditation.

The meaning of Upaasana

Previously we defined concentration as the intellect's ability to direct the mind and the power centres of perception and action to continue engaging in the present activity until it is completed, without allowing the mind to slip into the past or the future.

All people—even those with the weakest of intellects—have the power to concentrate for at least *some* amount of time. But that duration varies wildly from individual to individual.

The power of your concentration depends on the length of time that you can engage with a single thought, action, or task. And this, of course, does depend in large part on the relative strength (or weakness) of your intellect—its ability to keep your mind from letting your thoughts run helter-skelter. But in addition to having a strong intellect, the practice of Upaasana yoga can help you to systematically and gradually extend the amount of time you can hold your mind in its present activity. Your attention span widens.

The etymology of the term *Upaasana* makes this clear; *Upa* means 'near' while *Aasanaa* means 'to sit.' You may have also heard the term *Upavaasam,* which is a close equivalent of Upaasana; here, *Upa* again means 'near' while *Vaasam* means 'to dwell.'

[Additional Note: Upaasana yoga is in many ways synonymous with *Ashtaanga yoga* (sometimes referred to as *Samaadhi yoga*), a practice which has become highly commercialized. In this book, we will stick with the term Upaasana yoga.]

So, what exactly are we sitting or dwelling near when we practice this yoga?

Upaasana yoga consists of your intellect directing your mind to sit (or dwell) in thoughts of Brahman, the infinite reality. Doing so gradually deconditions your mind from its chaotic state, and prepares it for capturing infinity.

Even those with a strong intellect may not be able to sit or dwell in these thoughts for a long period at first. This can't be achieved overnight! However, *any person*—even someone with a weak intellect—can direct their mind toward thoughts of Brahman for at least a few minutes in a day. This is the purpose of Upaasana yoga—to begin training the mind to dwell near Brahman through preparatory meditation.

At first, it may be difficult since the mind is unruly. However, consider the difference you feel while concentrating on something that interests you versus something that does not interest you. When the latter happens, your mind is more likely to wander, and you must constantly exert your intellect to bring your mind back to the task at hand. But when you are interested in something or when you desire something deeply, your mind can focus easily, with minimal effort from the intellect. Herein your intense desire for Moksham and infinity comes into play, your Mu-Muk-Shatvam.

With Upaasana yoga, your mind will be unruly at first. But over time and with the help of the intellect, it will become much more willing to focus on dwelling near the thoughts of Brahman.

Before moving on to explore the practice of Upaasana yoga, let us briefly address an irony that highlights just how rare this practice has become.

An advanced practitioner who is capable of directing the mind toward thoughts of Brahman for the majority of the day would spend so much time concentrating that the person would not even have time to eat. This is how the term Upavassam wound up gaining its second, implied meaning i.e. 'fasting.' The irony is that the vast majority of Indians only understand the term Upavassam to mean 'fasting,' and are not aware of its primary meaning.

Upaasana yoga in practice

With that context, we are ready to discuss the practice of Upaasana yoga. By observing this practice, you will gradually and systematically be able to extend your ability to concentrate, starting from just a few minutes at a time.

Because this yoga is essentially a diluted form of Vedantic meditation (which we will address in the next chapter), we can consider it a type of preparatory or intermittent meditation—preparing one for Gnyaana yoga.

Readying oneself for Upaasana yoga

Before we begin, bear in mind the criticism of contemporary meditation that was presented in a previous chapter. Lengthy meditations that exceed one's capacity are *not at all* recommended as they are strenuous and detrimental to the mind, and do more harm than good. Do not be tempted to exceed your capacity when you are beginning. Instead, try to follow the recommendations below to build your capacity gradually. You must be attentive to your actual or current attention span, and not push yourself.

Now, in the first place, you must set aside time for this practice. You may start with as little as five to 15 minutes. The best time to perform these short meditations is early in the morning when your intellect is alert and your mind is relatively calm.

Next, choose a clean, quiet, and secluded location. Ensure that you are seated in a comfortable position, with folded legs, body erect, and eyes gently closed.

Thereafter, you must gradually withdraw all activities of the five organs of perception (*Gnyaana Indriya Nigraham*). Then, withdraw the activities of the five organs of action (*Karma Indriya Nigraham*).

In the beginning, the most difficult task will be to use your intellect to temporarily withdraw your mind from past worries and future anxieties, and to permit it to dwell only on the present (*Mano Nigraham*).

At this point, you will perform the regulated breathing exercise of Praanaayama. Do it for two to five minutes before beginning any of the meditations described below. Why is this important? Recall our previous discussions about the different components of the gross and subtle bodies—specifically, the power centres that act as the transition points between the two bodies. When your intellect directs your mind to focus on inhalation and exhalation, you are focusing on the vital energy layer, which has an impact on both the gross as well as the subtle body. This focus means that Praanaayama can calm the mind. Therefore, regulating your breathing in the direction of your intellect for a few minutes will bring your mind to a state more conducive to Upaasana yoga.

Once you have performed Praanaayama for a few minutes and brought your mind into that calm and balanced state, you will be ready to begin Upaasana yoga by observing one of the following preparatory meditations. Note that the following sections merely provide summaries of each type of meditation; a guided meditation source will be provided upon request via the email provided at the end of this book.

Relaxation meditation

As the name suggests, its objective is to relax your entire personality. You will start by relaxing your physical body. Mentally visualize all the physical parts of your body, starting from the head and gradually moving to the toe, relaxing each part of your body along the way. As you relax, you will initially feel the heavy presence of your body. This will gradually give way to a feeling of lightness, in which you feel completely withdrawn from your body and no longer feel its presence.

Once your body is relaxed, you will observe that your breathing has automatically become relaxed as well and is now calm and regulated. Observe your inhalation and exhalation, and continue to be relaxed.

Now comes the difficult task of relaxing the mind. This is difficult at first because of its natural tendency to slip into the past and future—largely on account of stress. The mind is a victim of stress in most individuals.

However, remember that stress is nothing but acute mental agitation caused by the inability to fulfil numerous desires at once. It is common for many desires to clamour for attention at the same time, all seeking instant gratification simultaneously. The pathetic state of your mind at this juncture is what we all know as *stress*.

While in the meditation posture, mentally suggest that you have developed all these impractical, stress-generating desires out of your ignorance. Suggest that by developing a stronger intellect, you will be able to eradicate those stressful desires at the thought level, without having to gratify them!

It will not be easy in the beginning. Your mind will still try to wander away within a few seconds. When that occurs, deploy your intellect and bring your mind back to the present. Do not chastise yourself, instead suggest that you can't control the past, only the present moment. Leave the result of the present moment to the future, surrender to the totality, and accept the future as it unfolds.

Though you are maintaining a relaxed, seated meditation posture while suggesting these things to yourself, it can help to direct your mind to visualize a relaxed environment—particularly a scenic, natural location that you are fond of. This can make it easier for the mind to relax and not wander.

Try to establish a consistent practice of relaxation meditation for five minutes at first, then gradually extend it to 10 or 15 fifteen minutes. Doing so will help reduce your stress level. If you are consistent in your efforts, this will help to relax your total personality even while you are not practicing this meditation.

Once you have succeeded in establishing a consistent practice, you will be qualified to begin the next type of preparatory meditation—concentration meditation.

Concentration meditation

As with the previous type of meditation, the name concentration meditation should make its purpose clear. The practice of concentration meditation aims to increase your attention span—the capacity of your intellect to direct your mind and body to continuously engage in the present activity until completion, without allowing the mind to slip into the past or the future.

There are three ways this type of mediation can be performed.

The first is **mental ritual**, or **mental worship**. With this approach, you can choose a deity from any religion you prefer. You then mentally perform any prescribed rituals related to that deity, following them from the beginning to the end without any distraction. You must perform all rituals *mentally*, corresponding to the prescribed ritual.

For example, if you are a devotee of Lord Shiva, you would sit in the meditation posture and then perform the exact procedural rituals *mentally* till the end without any distractions. This specific practice is called *Shiva Maanasa Pooja*.

The second option is **mental recitation**, in which you would mentally recite your favourite prayer. With this approach, you simply mentally recite it *without* delving into its meaning. Choose a prayer that would last 10 to 15 minutes, then mentally recite the prayer without any interruption until the completion of the prayer in its entirety. This practice is called *Maanasa Paaraayanam*.

The third option is the **mental chanting** of a powerful, pithy statement, called a *Mantram*. This is a statement that pertains to Brahman the infinite since the purpose of Upaasana yoga is to refine your mind for the goal of pursuing Brahman. Naturally, you would choose a mantram related to Brahman. For example, you might select *"Om Namah Shivaayam,"* and chant it mentally. This practice is called *Maanasa Japam*.

Depending on the amount of time you can concentrate, you can perform the mental worship, recitation, or chanting alone or in sequence—one after the other.

A consistent practice of both relaxation and concentration meditation will refine your mind and increase its subtlety, preparing you for the next type—macrocosmic meditation.

Macrocosmic meditation

While concentration meditation invites you to focus on the ritual worship of deities and the recitation of prayers or mantram, macrocosmic meditation focuses on an infinitely greater scope—the full spectrum of the cosmos.

While seated in the meditation posture, you will begin by visualizing the entire cosmos—starting from the minutest of the subatomic particles, you will work your way up to microorganisms, plants, animals, your existence in the vast expanse, the natural world around you, the country you live in, planet Earth, our solar system, the Milky Way galaxy, local cluster galaxies, and finally, the entire observable universe.

You will then visualize the magnitude of infinite Brahman and consciousness as not only pervading the entire observable universe but also *beyond* it!

Next, you will visualize that your personality is an infinitesimal tiny spec in the entire observable universe and that the entire observable universe is also an infinitesimal tiny spec within the all-pervading, infinite Brahman.

Finally, you will visualise how your sincere pursuit will ultimately catapult you to become *one* with the infinite Brahman.

Macrocosmic meditation, termed *Vishwa Roopa Eeshwara Dhyaanam*, will drastically change the way you view yourself, your problems and the world. Rather than seeing yourself as a separate entity, you will identify the oneness in all the objects and beings of the world! This refinement will reward you with an expanded, resilient mind.

As you gradually increase your capacity, you can eventually practice this meditation for 20 to 30 minutes a day. Once you have developed the ability to do so, you will be prepared for the fourth type—Transformative meditation.

Transformative meditation

No matter how meticulously you perform the previous meditations, if your inner personality is infested with negative thought patterns and negative desires then all of your effort will remain futile.

As human beings, we have complex personalities—a mixed bag of good and bad. And our awareness of that is inescapable. For those with a strong intellect, their flaws will stand out clearly. For those with a weaker intellect, their conscience will suggest what is wrong within.

Transformation meditation focuses on mentally assessing one's inner personality. While practicing this meditation, you identify and ignore negative aspects of your personality, no longer nurturing them with a further flow of thought. Your thoughts are the building blocks of your mind—as you think so you shall become. By depriving them, you can begin eliminating them from your personality.

Instead, you will meditate on the thought that only a refined, positive mind can enjoy peace and happiness. You meditate on the thought that material possessions will not guarantee happiness, and that, even without them, the mind can be at peace.

Thoughts have a tremendous capability to make things real. This power is called *Sankalpa Shakti*. In this meditative posture, you tap into this power. By consistently entertaining powerful, positive thoughts and visualizing that you are a transformed person, you will gradually change your thought pattern. In due course, this shift will gradually transform your personality.

These four preparatory meditations on various higher values collectively constitute Upaasana yoga. Even though they can be performed for only a few minutes each day, when you practice them consistently and systematically over some time you will gain six-fold wealth (Shatka Sampathi), which qualifies you to take up Gnyaana yoga.

One last concluding note: There is no hierarchy in practicing Karma yoga and Upaasana yoga; that is, you need not 'complete' one before practicing the other. Ideally, you *will* have established sufficient practice in both before entering Gnyaana yoga. However, for those who are curious, it is not wrong to explore the initial stages of Gnyaana yoga while you are still progressing with Karma and Upaasana yogas.

Chapter 27

Gnyaana Yoga

Once you have established a disciplined practice of Karma yoga and Upaasana yoga and have attained all the necessary prerequisites, you will finally be ready to apply yourself to Gnyaana yoga—the path of knowledge.

Gnyaana yoga is a rigorous method for gaining knowledge of the infinite Brahman, which can also be described as knowledge of the self, or knowledge of the "I" in you. This is represented by Moksham, the fourth human goal, which brings absolute freedom from bondage and dependence on the beings and objects of the world.

The primary cause of that bondage and dependence (and associated suffering, stress, and agitation) are the other three human goals—physical security, material comfort, and invisible wealth. Because these three goals inevitably involve beings and/or objects of the world, their pursuit will always involve impermanence, suffering, and a lack of fulfilment. And, what's more, your dependence on them is not only tied to *this* life but is rather the accumulation of *countless* past births.

In attaining the fourth Purushartham, you will become one with the infinite Brahman and enter a state in which you will no longer be affected by the presence or absence of any objects or beings of the world. Instead, you will effortlessly accept any unchangeable situation without any fear. You will be able to act decisively and with ease in any aspect of your life that must be controlled and changed, guided by a strong intellect that can easily identify what can and cannot be changed.

This refined state can only be achieved through a deliberate, attentive practice of Gnyaana yoga.

The logic and practice of Gnyaana yoga

If the knowledge of the self and infinite consciousness brings about liberation from bondage, then it's natural to wonder how one goes about *obtaining* that knowledge.

To obtain any type of knowledge in this world, one needs access to some source for that knowledge (*Pramaanam*).

When it comes to objective knowledge, you will generally rely on two sources:

- Sensory perceptions: primary source of objective knowledge.
- Inferences (such as postulations, comparisons, and so on): secondary source of objective knowledge.

Through our perceptions and inferences, we are capable of obtaining a massive amount of knowledge about the objective world, from the nature of the minutest subatomic particles up to the scale of the ever-expanding observable universe!

However great the power of human perception and inference, it does come with a limitation. While these powers can help us gain a great deal of knowledge about the objective universe, they cannot directly help with knowledge of the self, the pure consciousness, the real subject within that enables all the perceptions and inferences.

An example will make this concept clearer. While your eyes may be capable of seeing all perceptible objects of the world, they are *not* capable of seeing themselves directly. The only way they can see themselves is if some external medium reflects them, for example, if you look in a mirror.

Knowledge of the self is also subjective since a mirror is required to gain that knowledge. That mirror is the Vedantic scripture—namely, the Bhagavad Gita, Upanishads, and Brahma Sutras, the canonical triad we have mentioned previously.

Gnyaana yoga in practice

These Vedantic scriptures need to be studied systematically and consistently in order to gain knowledge of the self. And that systematic study, reflection, and assimilation is called Gnyaana yoga.

It comprises three stages:

- **Shravanam**, or learning the Vedantic scriptures
- **Mananam**, or reflecting on the Vedantic scriptures
- **Nidhi-Dhyaa-Sanam**, or Vedantic meditation

We will spend the remainder of this chapter discussing these stages in order.

Stage one: Shravanam or learning the Vedantic scriptures

The first stage of Gnyaana yoga involves the systematic study of the Vedantic scriptures. Though these texts are widely available and have been translated into many languages, Shravanam is not merely a matter of picking up a book and starting to read. Because this study involves rarefied concepts involving the place of the self within the infinite, you will need a competent teacher or guru, to guide you.

Consider some highly complex job, like that of a fighter pilot. A person cannot just *become* one. The person must undergo extensive, systematic training, working closely with qualified instructors (who have gone through *even more* training themselves). Only after a long period of study and practice will the trainee be qualified to perform the job. But one cannot do it without the guidance of the instructors.

Similarly, Shravanam involves the systematic, sustained study of Vedantic scriptures for a substantial length of time, under the expert guidance of a competent guru.

During this stage, your guru will prescribe a carefully planned schedule for studying the texts. It is highly recommended that you stick to the schedule and avoid the temptation to deviate. You will likely have many doubts and questions about the scriptures. However, you should resist your natural urge to seek clarification and get immediate answers.

Why? Because many Vedantic scriptures are conversational, featuring dialogues between a qualified student and a competent teacher. Thus, the texts will address many of the doubts and questions that you are likely to have and will respond to them in due course as you progress with your studies.

Of course, if you are a highly intellectually curious person, there is nothing wrong with posing questions to your guru during the process. But you should know that the Vedantic scriptures will eventually answer many of your questions, as they cover the subject matter comprehensively, from every point of view.

Your study will primarily involve the canonical triad, though your guru may also prescribe various other Vedantic texts as well. Based on your intellectual capacity, your guru will select *only* the necessary Vedantic texts that are sufficient for you to attain your goals (Purushartham).

These texts will address the exact concepts we have been discussing throughout this book: human goals, human composition, consciousness, creation, and so forth. This book is drawn directly from Vedanta; in a sense, it has provided you with a glimpse as to what the practice of Shravanam will be like!

The scriptures delve into the following six factors:

- You, the individual
- The universe
- Infinity, God, or Brahman
- The relationship between you, the universe, and God
- Moksham or salvation
- Methods (yogas) for gaining Moksham

Thus, the practice of Shravanam will remove your ignorance, preparing you for the next stage of Gnyaana yoga.

Stage two: Mananam or reflecting on the Vedantic scriptures

Once you have completed the systematic study of the Vedantic texts prescribed by your guru, you will move on to the second stage of Gnyaana yoga, which is known as *Mananam.*

While the process of Shravanam will address most of your doubts and questions, there will likely still remain some points on which you seek clarification. During Mananam, you are encouraged to think for yourself, applying your intellect to the available information to reflect on the concepts you've learned.

In this way, you will be able to derive your answers to your outstanding doubts and questions.

If you are still unable to get clarity on any concept, you can certainly raise your concerns with your guru who will help you understand. However, you are encouraged to use your intellect to build your understanding during this stage since it is the best way to remove any remaining doubts.

The outcome of Mananam, the second stage of Gnyaana yoga, is that you can remove all doubt from your knowledge of Vedanta. Vedantic knowledge will become your conviction, which is the prerequisite for the third stage of this practice.

Stage three: Nidhi-Dhyaa-Sanam or Vedantic meditation

Having gained sufficient knowledge and become thoroughly convinced of the contents of the Vedantic texts, you will be ready to take the next important step in the pursuit of infinity—Vedantic meditation (Nidhi-Dhyaa-Sanam).

Vedantic meditation is the process of assimilation of the Vedantic knowledge you had gained through Shravanam and Mananam. This assimilatory process will be far easier if you have properly performed Karma yoga and Upaasana yoga since those practices help to cultivate the mind and intellect for this advanced undertaking.

Vedantic meditation helps you bridge the gap between what you know of Vedanta and who you *are* now, in the present. Your personality, of course, is the result of your past experiences, and it may be a mixed bag of both good and bad qualities. You may be a victim of your emotions, suffering the consequences of that state. You may be deeply conditioned to tolerate suffering and attachment to the temporary.

Through the process of Vedantic meditation, you will gradually decondition yourself from this way of being. In time, you will become what you know of Vedantic knowledge—you will *become* Vedantic wisdom! In doing so, you will become a *Gnyaana Nishtaa*, a person who is firmly established in Vedantic wisdom and who has gained complete emotional strength. There are a few key components involved in this transformation.

First, you must continue your study of the Vedantic texts. You must also try to limit your associations and interactions to fellow *Vedantins* only (to the extent possible). This practice is known as *Sat-Sangam*; you must try to maintain it throughout the practice of Gnyaana yoga.

In doing so, you will ensure that the stimuli you receive from the external world are in line with a Vedantic way of life. This limits opportunities for your past conditioning to rise to the surface, which is an important part of *deconditioning* yourself from those behaviours. Over time, you will no longer react to your old habits, even when you receive the same stimuli. As a result, your intellect will become more and more available to you. Eventually, it will be available at all times—before you act, while you act, and after you act.

As such, your actions will be governed by Vedantic teaching and not by your past conditioning. You will be able to invoke Vedantic knowledge anytime, anywhere, and put it into practice effortlessly, reflexively, and instantaneously.

Recall the various preparatory meditations involved in Upaasana yoga, wherein you train your intellect to bring your mind back to the task at hand during short meditation sessions.

In Vedantic meditation, most of your day will be dedicated to the contemplation of Vedantic principles. With a purified mind and intellect, you will gradually find it easier to sustain your concentration since it will take less effort to keep your mind and your intellect on the same page. Your mind will readily and meticulously follow the direction of your intellect, while your intellect will clearly recognize that your mind's value systems are precisely aligned with it.

Eventually, you will be left with very few thoughts about the world around you, and most of your day will be spent reflecting on the core teachings of Vedanta. In other words, you will become meditative and your entire waking life will be devoted to Vedantic meditation.

This state represents the end of the three-part progression of Gnyaana yoga. In the first stage, Shravanam, you depend on your guru's direction; in the second stage, Mananam, you rely on your intellect, but still have the support of your guru; and in the third stage, Nidhi-Dhyaa-Sanam, you are totally self-reliant.

By that time, you will be completely occupied with the practice of **Vedantic meditation—the capacity to direct your mind to a single thought toward the infinite Brahman, under the continuous observation of your intellect and to the exclusion of all other thoughts, for a sustained period of time.**

So, what then? What happens to your personality once you have achieved this rarefied state of evolution? We will discuss this subject in the following chapter.

Chapter 28

You Merge with Infinity

This chapter presents the central premise of the entire philosophy of Vedanta—merging with (or capturing) infinity. Every theme and concept we have discussed throughout the book has led up to this point. Now, we will address the change that overcomes you as you reach and *achieve* this state, which can only occur once you have evolved and are reveling in an advanced meditative lifestyle—Nidhi-Dhyaa-Sanam, the third stage of Gnyaana yoga.

If you observe a diligent Vedantic practice, you will gradually come to spend the majority of each day immersed in thoughts of the infinite Brahman. This state is known as *Nithya Anithya Viveka Vichaaram*—"constant dwelling in thoughts of the permanent, infinite Brahman, and negating the impermanent aspects of the world." Only when you have attained this state will you finally be able to merge with the infinite.

As we proceed through this chapter, we will be referring to concepts presented in the chapters on consciousness and creation. As such, you may wish to take a moment to turn back to those chapters and refresh your memory before proceeding. Returning to those earlier concepts in conjunction with the following pages will help you develop a deeper understanding of the ideas presented in this chapter.

Macrocosmic and microcosmic reflections of the infinite

Earlier in this book we detailed the process of creation. Specifically, we illustrated how creation unfolded in two stages. First, the causal universe gave rise to the five fundamental states of subtle matter, the subtle universe, and the subtle body which in turn, gave rise to the five fundamental states of gross matter, the gross universe, and the gross body.

The causal, subtle, and gross universes form a single combined entity known as the observable universe, also known as the macrocosm or *Brahmaandam*.

Similarly, the causal, subtle, and gross bodies combine to form a single entity as well, known as the individual or the microcosm or *Pindaandam*.

When the macrocosmic universe (Eeshwaraa) and the microcosmic individual (*Jeevaa*) merge (*Aikyam*), that 'event' is known as *Jeevaa Eeshwara Aikyam* or 'You merge with the universe.'

Macrocosm and microcosm defined

Before we can address the nature of that merger, we'll first need to understand what aspects in both the macrocosmic universe and the microcosmic individual are getting merged!

Out of the three components of the macrocosm given below

- The causal universe (*Kaarana Prapancham*),
- The subtle universe (*Shookshma Prapancham*), and
- The gross universe (*Sthoola Prapancham*),

both the subtle and gross universes begin with the Big Bang (*Srishti*) and end in the Big Crunch or Big Rip (*Pra-Layam*), while the causal universe is timeless and has no beginning (*Anaadhi*).

Similarly, out of the three components of the microcosm given below

- The causal body (*Kaarana Shareeram*),
- The subtle body (*Shookshma Shareeram*), and
- The gross body (*Sthoola Shareeram*),

the subtle and gross bodies begin in the Big Bang (Srishti) and end in the Big Crunch or Big Rip (Pra-Layam), while the causal body is timeless and has no beginning (Anaadhi).

However, despite the differences between each of these six components, one thing they have in common is that they are composed of *matter*; whether it be causal, subtle, or gross, being *material* in nature is a trait they all share.

Macrocosm and microcosm: Attributes of the attribute-less infinite

Each of these six components is also pervaded by consciousness. Being fundamentally matter, they comprise living beings and inert objects that manifest consciousness. All living beings by possessing the subtle body are conscious. All the inert objects, by their mere existence, express consciousness.

The existence of any inert object is proved, by cognizance of that object by a thought. And for a thought to be expressed, it must borrow consciousness from original consciousness. Therefore, the SIX matter components act as a medium for the expression of the all-pervading, absolute, pure, infinite, original CONSCIOUSNESS!

This infinite consciousness (known as *Nirguna Brahman*) is totally devoid of attributes. However, Nirguna Brahman *appears to gain* attributes when it is manifested through one of the six components of the macrocosm and microcosm. In a sense, these six material components act as reflecting media, through which the infinite consciousness expresses itself in various ways.

For those with more of an academic interest in the subject, another Sanskrit term for the infinite pure consciousness is *Bimba Chaitanyam;* the term for *reflected* consciousness through the six reflecting media is *Prathi Bimba Chaitanyam.*

How can it be that the infinite consciousness is attribute-less, while reflected consciousness *does* have attributes? You can liken this phenomenon to a carnival house of mirrors. As you look into each of the different mirrors—a concave mirror, a convex mirror, a corrugated mirror, and so on—you will see the same original entity (your physical form) reflected as if it possessed various attributes which do not actually exist.

Similarly, each of the six material components of the macrocosm and microcosm reflects a different distorted expression of the infinite consciousness, lending it a unique set of attributes in the process.

Vedanta describes the macrocosmic reflections of the infinite consciousness as follows:

- The infinite consciousness reflected through the causal universe is known as *Antaryaamee*
- The infinite consciousness reflected through the subtle universe is known as *Hiranya-Garbhaa*
- The infinite consciousness reflected through the gross universe is known as *Viraat*

Remember, these universes form an inseparable combined entity, Brahmaandam or the macrocosmic universe.

That aspect of the ***pure consciousness reflected through this combined entity Brahmaandam or macrocosmic universe is known as Eeshwaraa.***

Vedanta describes the microcosmic reflections of the infinite consciousness as follows:

- That reflected through the causal body is known as *Praagnya*
- That reflected through the subtle body is known as *Taijasaa*
- That reflected through the gross body is known as *Vishwaa*

These bodies also form an inseparable combined entity, the microcosmic individual (Pindaandam). *That aspect of the **pure consciousness reflected through this combined entity is known as Jeevaa.***

Despite gaining names and apparent attributes when reflected through the various components of the macrocosm and microcosm, it is essential to remember that the infinite consciousness *itself* lacks attributes of its own. It is only through reflection that it *appears* to exhibit certain attributes, which are truly manifestations of the infinite consciousness *through* various types of matter.

Therefore, JEEVAA is microcosmic reflected consciousness and EESHWARAA is macrocosmic reflected consciousness.

It is pertinent now to recognize that the infinite, pure, original consciousness, brahman, is all by itself, self-effulgent and it is UNDISTORTED.

Attributes of Jeevaa

Compared with the attributes of the macrocosmic reflected consciousness (Eeshwaraa), the attributes of the microcosmic reflected consciousness (Jeevaa) are considered inferior.

This is on account of numerous factors, including the limited pervasiveness of the causal, subtle, and gross bodies; their exposure to external factors such as weather, natural calamities, and other physical forces; their physical limitations, specifically concerning the sense organs; their susceptibility to emotion and the limitations of knowledge; and their subjection to all the other various and constant changes that occur throughout life.

Among individuals, there is a gradation of inferiority and superiority in these attributes; however, compared with the macrocosm, the attributes of *all* individuals are vastly inferior.

Collectively, the inferior attributes of Jeevaa are called *Nikrushta Guna.*

Attributes of Eeshwaraa

Compared to the attributes of the microcosmic reflected consciousness, the attributes of the macrocosmic reflected consciousness are considered superior.

This is on account of numerous factors, including the all-pervasiveness of time, space, and causation, which go on endlessly and pervade the entire observable and unobservable universe (in addition to the multiverses); the perfect order that extends throughout the planetary systems, galaxies, and the universe; the freedom from limitation by the mind and intellect; and so on.

Collectively, the superior attributes of Eeshwaraa are known as *Uthkrushta Guna.*

The infinite: The attribute-less attribute of the macrocosm and microcosm

Thus, Eeshwaraa is with superior attributes and Jeevaa is with inferior attributes. The reflecting media cause the apparent differences in attributes.

In other words, if you can visualize and remove the six reflecting mediums from all the reflected consciousnesses, you will then be able to deduce that what remains is the original consciousness in its own state.

So, when the Jeevaa is viewed through the reflecting mediums of gross, subtle, causal bodies, it appears inferior. When Eeshwaraa is viewed through reflecting mediums gross, subtle, causal universes, it appears superior.

If both Jeevaa and Eeshwaraa are viewed from their essential intrinsic nature, both are the infinite, original consciousness or Brahman.

Put another way, while the differences between Eeshwaraa and Jeevaa may *appear* to be substantial, they are really only superficial and incidental. **If you are capable of *removing* those superficial, reflective attributes from the macrocosm and microcosm, that which remains can only be the infinite consciousness in its own state.**

When you consider the macrocosm and microcosm *through their attributes*, they will appear to be relatively superior and inferior. But when you consider them *through their essential, intrinsic natures*, they are identical—they are both the same with the infinite, original consciousness or Brahman.

Vedanta provides a wonderfully illustrative example which drives this concept home.

Consider the relationship between the ocean and a wave. These terms clearly refer to two distinct entities; one would never mistake one for the other. Waves, too, are quite different from one another. One wave may be a towering tidal wave, while another may be a barely noticeable wavelet. However, compared with the ocean, both are trivial—short-lived, small in scale, and one of many. In this respect, a wave is the equivalent of the microcosmic Jeevaa, while the ocean is the equivalent of the macrocosmic Eeshwaraa.

And yet, despite their differences, both the wave and the ocean are essentially the same. They are nothing but water. Their names and forms may give them differing attributes, but their intrinsic nature is identical. In fact, there is almost no such thing as 'wave' and 'ocean,' in essence, there is only water.

This is precisely the relationship between the macrocosm, the microcosm, and the infinite consciousness.

Merging with the infinite

This is precisely the understanding that you will be prepared to attain (on a deep, fundamental level) once you have reached the final stage of Gnyaana yoga, Nidhi-Dhyaa-Sanam.

By the time you complete your systematic study of the prescribed Vedantic texts under a guru during the process of Shravanam, you will have reflected on the texts for an extended period of time during the process of Mananam.

After spending your days in the advanced meditative state of Nidhi-Dhyaa-Sanam, you will gradually realize that, like the wave and ocean, your own Jeevaa shares the same intrinsic nature as Eeshwaraa—and that both are mere reflections of the original, infinite consciousness.

As a result of your dedication to your spiritual journey, your unswerving conviction, rigorous cognition, and total identification with the essential oneness of original, infinite consciousness, you will have finally arrived at the point where you are capable of merging with infinity (*Jeevaa Eeshwara Aikyam*). In this state, you become, a *Gnyaana-Nishtaa,* totally established with the wisdom of the infinite pure consciousness.

Many chapters ago, during our discussion of the logic behind capturing infinity, we introduced the happiness formula. You may wish to turn back to that chapter now to refresh your memory—although, having made it this far, you may already perceive how that idea fits in here, and how it is converging with our discussion of the practice of Karma, Upaasana, and Gnyaana yogas.

As you develop and enhance your yogic practice, you will gradually reduce the number of unfulfilled desires. As you reduce the denominator of the happiness formula, your happiness will increase. But, more significantly, you will come closer and closer to the point at which you will entertain only a single remaining thought—the thought of capturing infinity.

In Nidhi-Dhyaa-Sanam, you will finally achieve this state of total absorption in a single thought of identification with the original, infinite consciousness. And the longer you remain in this state, the more and more attenuated your thoughts will become. Until like a thin steel wire stretched to its snapping point, your last thought will finally snap, enabling the denominator to become zero.

At that point, you will attain infinite happiness and be catapulted into the infinite consciousness. You, the Gnyaana Nishtaa, will become liberated and will lose your individuality as you become one with the infinite Brahman—as you *become* Brahman. **While you were once an infinitesimal dot in the universe, now the entire universe will become a tiny dot within YOU!**

Having, at last, attained that state through diligent practice of Gnyaana yoga, you will finally come to experience what is described in Vedanta as *Gnyaana Phalam,* or the 'ultimate benefit' of becoming and being a Gnyaana Nishtaa.

This ultimate benefit, termed Moksham or Mukthi consists of two parts: first *Jeevan Mukthi,* and then *Vidhaeha Mukthi.*

Jeevan Mukthi is the experience of liberation (Moksham) while you are alive. It is the experience of total salvation within this lifetime. When you come to the end of your physical life as a *Jeevan Mukthaa* and finally die, you will attain *Vidhaeha Mukthi*—permanent, uninterrupted, Infinite liberation after death. You have broken the endless cycle of birth and death, and are thereafter totally free from rebirth (*Punar-Jenmam*).

There is more to explore in order to fully understand Jeevan Mukthi and Vidhaeha Mukthi. In the next chapter, we will discuss the Vedantic concept of the law of past actions (that is, Karma), and what that law means for the *Gnyaana Nishtaa*—both in this lifetime and after the moment of their physical death.

Chapter 29

The Law of Karma: The Never-Ending Story... Ends

Your diligent practice of Gnyaana yoga will eventually culminate when you achieve a sustained meditative state and, finally, succeed in capturing and merging with infinity.

When this occurs, you will become known as a Gnyaana Nishtaa and will attain two types of liberation in the order of: Jeevan Mukthi, or liberation in life; and Vidhaeha Mukthi, or Infinite liberation after death.

Let us now explore how these two concepts fit together, and how they relate to the Law of Karma—the logic that governs and ends the seemingly endless cycle of birth and death.

Life and death of a Gnyaana Nishtaa

Before we can fully approach the topic of the Law of Karma, it's essential to understand the experience that awaits the Gnyaana Nishtaa who has succeeded in capturing infinity. Having achieved this rare feat, the Gnyaana Nishtaa will have been liberated from the never-ending story of birth, death, and rebirth.

While the Law of Karma governs the cycle of reincarnation, it will be helpful to have more perspective on how that cycle *stops*. This will provide useful context when we finally turn to the logic of the law. With that in mind, we will first discuss the nature of Jeevan Mukthi, or liberation in this lifetime.

Jeevan Mukthi: Salvation in life

As a Jeevan Mukthaa, you will achieve total independence at all levels of your personality.

At the physical level, the presence or absence of any object, being, or other physical thing will not affect your equanimity. Of course, your physical body will still experience pleasure and pain depending on circumstance. However, you will be absolutely detached from all bodily comforts and discomforts. Your level of happiness will be the same regardless of whether you live in a palace or on the streets.

At the emotional level, you will be similarly free from dependency. External factors will no longer affect your mental equilibrium, and you will no longer experience internal emotions that cause mental instability. While others around you may create and be swept up by emotional situations, such things will no longer have an impact on you. You will remain completely objective and maintain perfect mental equanimity under all circumstances.

At the intellectual level, you will also be unaffected by internal and external destabilization. Even if others criticize, censure, or make other attempts to dishonor you, your intellectual composure will not be affected. You will remain the embodiment of independence (*Swathanthram*), enjoying absolute fulfillment (*Poornathwam*) in this life.

Your equanimity will be based on the understanding that all events and changes in this world only affect the six reflecting media, and are only perceived and experienced by reflected consciousness. By this point, however, you have evolved far beyond reflections. You have fully identified and become one with the original, infinite consciousness. Any changes that pertain to the six reflecting media and consciousness are no concern of yours.

Having attained Jeevan Mukthi, you will remain in this state of infinite happiness, endless fulfillment, and complete independence until the day you die—at which point, you will attain Vidhaeha Mukthi, liberation after physical death.

Vidhaeha Mukthi: Freedom from rebirth

What, exactly, are you liberated *from* at the point of death? From the cycle of rebirth, or *Punar-Jenmam.* Rebirth, and freedom from it are governed by the Law of Karma as described in the Vedantic texts. (*Karma,* in this usage, means 'action.')

To understand how the Law of Karma works and how capturing infinity can help you break it, we'll need to recall certain principles you are already familiar with—both from your independent study and earlier discussions in this book.

Specifically, you may wish to refresh your understanding of the concept of invisible wealth (also known as good deeds or Punnyam), one of the three categories of human goals. Certain laws of physics, such as Newton's laws of motion and the law of conservation of matter and energy, will also help illustrate the nature of the Law of Karma.

Invisible deeds: What are they and how are they formed

Any deliberate action (Karma) you take in this life will produce two kinds of effects—visible effects (*Dhrishta-Phalam*) and invisible effects (*Adhrishta-Phalam*). To clarify how these terms are derived, *Dhrishta* means 'visible,' while *Adhrishta* means 'invisible' and *Phalam* means 'effect.'

To illustrate the meaning behind these terms, imagine that you performed some charitable action such as helping those in need to obtain food, clothing, and shelter. This would obviously entail some reduction of your material resources while increasing the material resources and the well-being of others. Those physical changes are known as Dhrishta-Phalam or visible effects.

In addition to these physical changes, your actions also produce corresponding invisible effects (Adhrishta-Phalam). In the above example, the invisible effect is that good deeds are added to the invisible wealth of your personality. These are known as Punnyam.

On the other hand, if you had intentionally performed a harmful, immoral, or otherwise ignoble action, your action would have produced negative invisible effects. Invisible bad deeds (*Paapam*) such as these are also added to the invisible ill-wealth of your personality, just as invisible good deeds are.

How is it determined whether an action would produce Punnyam or Paapam? It all depends on intentionality and conscience. Cause and effect can be complex, and your actions don't always produce the results you intended. But **the invisible effect of your Karma is the result of your intent while carrying it out.**

Most people know right from wrong. Any person with a reasonably well-developed intellect will also have a clear sense of which actions will produce Punnyam, and which will produce Paapam. And even an immoral person has some understanding of the difference between right and wrong. However, their strong negative intentions generally cloud their conscience, leading them to continue to take actions that produce negative effects.

Most of the world's belief systems have issued their guidelines for the right ways of living, whether they be in the form of commandments, injunctions, or other collections of 'dos and don'ts.' However, some of these injunctions may be open to interpretation, while others may have once been applicable in the distant past but have lost their meaning in modern times.

Thus, while such belief systems can help inform the right behavior, it is ultimately up to each person to act according to their true convictions. And it is the intention behind those convictions and actions that determine the invisible effects of your behavior in this world.

Invisible deeds lead to rebirth

As you go through your life taking one action after another, you will continue accumulating good deeds (Punnyams) and bad deeds (Paapams). These invisible effects generally remain invisible in your personality for a period of time before eventually finding a physical manifestation. Punnyams will manifest as visible happiness (*Sukham*), while Paapams will manifest as visible pain (*Dukkham*).

However, it must be noted that there is no direct correlation between the actions you take and the physical manifestations you will ultimately experience in this life. Why? Consider the law of resultant force, which describes how objects move when multiple forces act on them. When multiple forces coming from multiple directions all exert force on the same object, the result is that those forces combine to exert a net force that is sometimes quite different from any of the individual forces themselves.

Similarly, Punnyams and Paapams of different magnitudes will combine to affect your life in unpredictable ways. What's more, you may not experience their physical manifestations during your life. When you reach the point of physical death, your personality may still contain numerous unrealized invisible deeds. And on top of that, your personality will almost certainly contain numerous unfulfilled desires that you accumulated throughout your life as well.

These desires and invisible deeds reside in your subtle body, in the form of thoughts. Recall that thoughts are composed of subtle matter—meaning that they are material in nature. As the law of conservation of matter and energy dictates, matter and energy cannot be destroyed. This implies that, at the time of your death, the contents of your personality cannot simply vanish into thin air, they fold back to your subtle body and causal body. All the Punnyams, Paapams, and unfulfilled desires that remain in your personality *must* be conserved. How are they conserved? **They are conserved through your rebirth.**

Thus, your current birth (Jenma) is nothing more than the continuation of an endless chain of past millions of births and deaths, driven by the Punnyams, Paapams, and unfulfilled desires of your personality at the time of each death. This endless chain of reincarnation (Punar-Jenmam) will continue endlessly as long as you continue to accumulate new Punnyams, Paapams, and desires.

It's a never-ending story—*unless* you make an effort to bring that story to an end by seeking and capturing infinity!

It is in Vidhaeha Mukthi that you will finally succeed in liberating yourself from this cycle of reincarnation, which is governed by the Law of Karma.

Understanding the Law of Karma

The Law of Karma accounts for all the various experiences of all living things in this world.

It explains the disparities in the circumstances of one's birth, that is, why some are born to rich families, while others are born into poverty, with physical infirmities or with other conditions that make life difficult.

It explains why wars, natural calamities, and diseases can wreak havoc on the lives of some while sparing others. It explains why we share the world with animals, plants, and other lower life forms. And it explains why (and how) each person inherited their unique mental temperament.

This last point is extremely important for one who aspires to capture infinity. Since the Law of Karma demonstrates how your personality in this life is the direct result of your experiences and choices in your past lives, the logical conclusion is that ***you* are solely responsible for your experience in this life and future lives; you are the only architect of your fortune and misfortune.** No external force or factor is responsible for your current state or your current happiness or unhappiness.

Recall that all events and changes in this world only affect the reflecting media, and are only perceived and experienced by the reflected consciousness. This demonstrates that even though you are pervaded by (and borrow consciousness from) the infinite consciousness, it is merely a witness to your actions. It is neither responsible for them nor affected by them. While many ignorant people attempt to blame God (the original consciousness) for the conditions of their lives, one who is acquainted with the Law of Karma will recognize the truth: that each person is responsible for their present situation.

Once you realize that your present is the cumulative result of your past actions across multiple lifetimes, the next realization is inevitable: your *future* situation depends on the actions you take today, tomorrow, and for the rest of your life.

Of course, it's important to maintain some perspective. Remember, your present situation is the result of *multiple lifetimes* of choices and actions. You won't be able to overcome that legacy and change your future overnight.

Say you had a room in your house that was painted a vibrant shade of red. If you wanted to paint the room green instead, would you immediately achieve the desired result after the first coat of green paint? No, you would need to apply several coats of green paint to finally obtain the shade of green you wanted.

Similarly, you may *want* to change the conditions of your life, and may even begin to attempt to do so by performing the right actions from this moment forward. But you must remain realistic and recognize that you are working against the cumulative result of your past actions—an inheritance that will hang on to you for a while.

It is only through the diligent practice of Karma yoga, Upaasana yoga, and Gnyaana yoga that you will gradually be able to neutralize and overcome that legacy of past action.

Technical definition of the Law of Karma

A technical definition of the Law of Karma may help to solidify this concept. All the Punnyams, Paapams, and unfulfilled desires that have accumulated over past endless births (Jenmams) are collectively known as *Sanchitham*. With each new birth, a small subset of your Sanchitham will emerge from the broader reservoir. This small subset is known as Praa-Rabdham, which refers to the Punnyams, Paapams, and desires that will seek manifestation and fulfillment in your new birth.

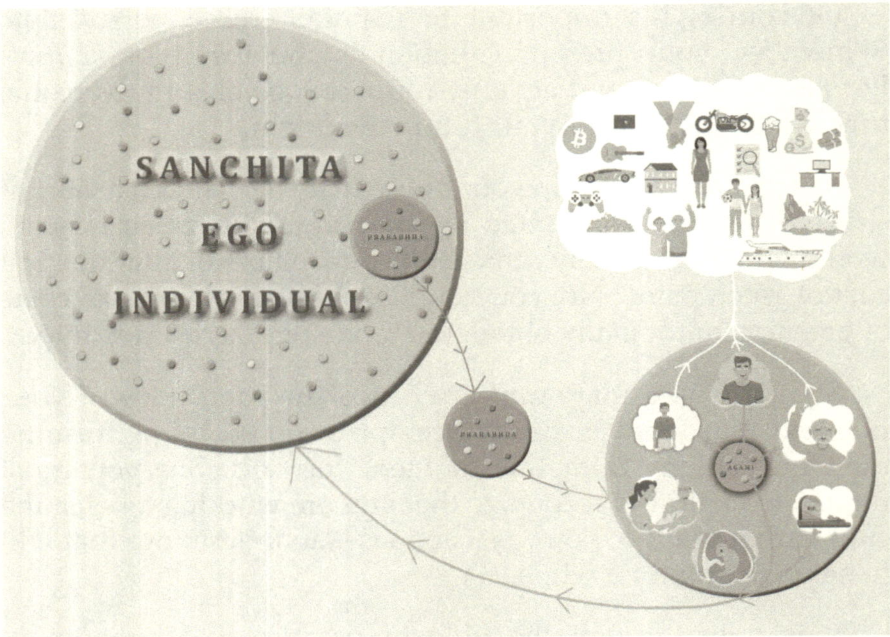

When you are born, your full Sanchitham will remain in the causal body and will have no further effect on your experience of life. When you take birth, you are *born* only with your Praa-Rabdham. This Praa-Rabdham determines the conditions of your birth—rich or poor; human, animal, or plant; and so forth.

Your Praa-Rabdham has a set of Punnyams, Paapams, a set of unfulfilled desires from past life, and these sets can only be exhausted through life's experience. However, at the same time as you are exhausting your Praa-Rabdham in the current life, you will *also* be accumulating brand new Punnyams, Paapams, and desires. This fresh set is known as Aagaamee.

When you have fully exhausted your Praa-Rabdham, you will also reach your physical death. At the time of death, you may have exhausted a *portion* of your Aagaamee, however, it is unlikely that you will have exhausted *all* of it. This remaining portion of your Aagaamee is then added to your Sanchitham—

which once again forms the basis for a *new* Praa-Rabdham which will catapult you into your next birth.

This endless cycle of Sanchitham, Praa-Rabdham, and Aagaamee is what drives the cycle of reincarnation. It is what brought you into this life, right up to the present moment!

Before we move on to put this technical definition in the context of ideas we have already discussed, let us make one passing observation about lower forms of life, such as animals and plants. When an individual is reincarnated as such a life form, they are indeed born with a Praa-Rabdham, just like a human. However, because they are born without an intellect, they are unable to perform intentional acts. Since intentionality is the basis for obtaining Punnyams and Paapams, therefore animals and plants will not gain any Aagaamee during their lives. They will simply live until they have exhausted their Praa-Rabdham and then die, only to be born once again—hopefully as a human.

The Law of Karma in context

Let us now return to the concept of Vidhaeha Mukthi. Your enhanced awareness of the Law of Karma will help refine your understanding of this important principle even further than before.

After you have achieved the sustained meditative state known as Nidhi-Dhyaa-Sanam and have finally captured and merged with infinity, you will become a Gnyaana-Nishtaa and attain Jeevan Mukthi—salvation in this life. In doing so, you will gain ultimate knowledge of the original, infinite consciousness, also known as Brahman or God.

Consider what the act of gaining knowledge actually *does*, in relative terms. When you gain knowledge of physics, you destroy the ignorance of physics. Likewise, **gaining knowledge of the infinite consciousness destroys your ignorance about it.**

There are a couple of corollaries we can derive from this statement.

First, identification with the reflecting media and reflected consciousness are byproducts of ignorance of the infinite consciousness. And Sanchitham all along always belonged *only* to the reflecting media and the reflecting consciousness. This means that, when you destroy your ignorance of the infinite consciousness, you in effect destroy your identification with any of its reflections.

Second, because your Sanchitham belongs to *your* individual personality, and your personality is merely the reflecting media and reflected consciousness, when you destroy your identification with those reflections and media you *also* destroy your Sanchitham.

Previously we said that, because thoughts are material, they cannot be destroyed—and that this phenomenon is ultimately responsible for the cycle of rebirth. If that is true, then how can Sanchitham be destroyed?

Vedanta offers the following explanation. Together, the reflecting media and reflected consciousness constitute the ego (*Aham-Kaaram*)—and this ego is the sense of belonging, that gives the Sanchitham an anchor. When Jeevan Mukthaa gains knowledge of the infinite consciousness, the ego is lost. This deprives the Sanchitham of its home and loses its locus standi. At that point, the Sanchitham simply loses its identification with the causal body and merges with the endless causal universe instead. In that respect, the Jeevan Mukthaa's Sanchitham is also 'destroyed.'

At this point, the Jeevan Mukthaa is no longer capable of performing actions with intentions. Therefore, Aagaamee is not accumulated any longer. Like anyone else, such a person will still retain the Praa-Rabdham present since birth, and experience both pleasant and unpleasant experiences for the remainder of the physical life.

However, because these experiences pertain to the reflecting media and the reflected consciousness alone, the Jeevan Mukthaa no longer identifies with them. These experiences are insignificant to one who is firmly ensconced in the infinite consciousness.

Thus, when the Jeevan Mukthaas exhaust their Praa-Rabdham, they will simply die as the Sanchitham is destroyed and Aagaamee is no longer being accumulated. Such persons will not be reborn into another physical body (Daeham) but will attain Vidhaeha Mukthi, liberation without a new body. That is to say, they will never be reborn again .

The never-ending story is thus brought to an end.

Chapter 30

What Next?

We have finally come to the end of our story. Looking back, we have covered quite a lot of territory: from the logic behind capturing infinity to the nuances of human composition and consciousness, and from the cyclical nature of the material universe to the yogic practices that can help you break free from the cycle of birth and rebirth.

At this point, it's only natural to wonder: What comes next? Now that you have obtained all of this knowledge, how should you integrate it into your life? What do you do once you put down this book?

If you have been able to comprehend even half of the ideas presented in these pages, you will likely know exactly what to do next. If you have made it this far, you are undoubtedly among the upper echelon of the rare 10 million intellects for whom this book was intended.

However, to make the path ahead even clearer, let us summarize some lessons you can take with you and start putting into practice *today*.

The path ahead

As you have surely gathered, the process of pursuing and capturing infinity does not happen overnight. It requires diligent, consistent application of the intellect—sometimes, over the course of more than one birth.

But, as with all journeys, your progress on the path toward the infinite begins with your first steps. Here are three steps that you can start taking right away.

First three steps

First, **begin performing the self-analysis that will help you discover the proportions of your mental temperament (Guna)**. This will be the starting point for understanding the right methods that will allow you to systematically progress toward a Saatvik Guna.

Second, start training **your physical body to prepare it for the rigors of the more advanced yogas you will practice later**. Start with basic Hatha yogas and simple Praanaayama breathing techniques.

Third, **try to set a relatively unselfish goal that will benefit others**. When you set such a goal and direct your thoughts and actions toward it, it will help enforce your commitment to self-development. By maintaining focus on such a goal, your mind's frivolous desires will fall by the wayside. It will also help prevent the formation of *new* unfulfilled desires, which will help you exhaust your Praa-Rabdham more efficiently.

Each of these steps will help to gradually prepare you for the practice of more advanced yogas—Upaasana yoga and Gnyaana yoga, which will ultimately allow you to decondition and purify your mind.

All along, your objective will be to reduce the gap between what you *know* and what you *are*. Guided by your robust intellect, you will eventually build robust mental resilience as well. This is the key to bridging the gap between your Vedantic knowledge and Vedantic wisdom—and, finally, *merging* with the Vedantic wisdom when you capture infinity.

Two common pitfalls

Of course, this is much easier said than done! Despite your commitment to your spiritual growth, you will face pitfalls that can challenge your progress toward Vedantic meditation (Nidhi-Dhyaa-Sanam), the culmination of Gnyaana yoga which leads to merging with the infinite consciousness.

Specifically, you will likely face challenges from two major sources: your five senses and your mind. These common pitfalls can confront you at any stage of your yogic practice. Thus, you must prepare yourself to recognize the risks and prevent them from obstructing your spiritual progress.

Controlling your senses

Your five sensory organs are constantly exposed to the material world. They are the pathways by which external stimuli enter your mind in the form of perceptions. And this constant bombardment by sensory perceptions has the dangerous potential to sway you from your path.

In order to *stay* on your path, you need to be prepared to practice *Indriya Nighra-ham*, or 'control of the five sense organs.'

To illustrate this practice, consider what happens to a tortoise when it senses danger—it withdraws its four legs and its head into its shell to protect itself. **You must condition yourself to respond to dangerous sensory perceptions just as the tortoise**

does. Be alert to the possibility that any of your sense organs may lure you away from your path at any time. Whenever this occurs, you must deploy your intellect and it **will withdraw your sense organs, removing them from the influence of external stimuli that may harm your spiritual progress.**

It is important to note that *control* over your sense organs does not imply *denial* of sensory experiences and sensory pleasures. You are free to responsibly enjoy the sensory pleasures of the world during your pursuit of infinity. However, you must rely on your intellect to ensure that this does not prevent you from losing focus on your goal.

Human history is littered with well-reputed individuals who have fallen prey to their inability to control their sense organs. You must be prepared to use your intellect to gain mastery over your sense organs and avoid becoming the servant of external stimuli.

Controlling your mind

Your mind will likely be the most difficult obstacle you face during your pursuit of infinity.

Though you are constantly bombarded with sensory perceptions throughout your life, at least you can physically move away from potentially dangerous stimuli. This is not the case with your mind. You are stuck with it, no matter where you go or what you do.

Hence, you must prepare yourself to practice *Mano Nighraham*, or 'control of the mind.'

An example will help to illustrate the risks of an uncontrolled mind. Even if you have perfect control over your senses, you cannot stop external stimuli from entering your mind as perceptions. When this occurs, your mind may dwell on something you have perceived—for example, a person such as a friend or a loved one. This person may remain in your mind, even when you are not around them.

A few stray thoughts (*Vishaya Dhyaanam*) regarding this person will begin to stir in your mind at this point. **This is the key moment for your intellect to step in and control your mental activity. Since these thoughts are weak and are few in number, they can be easily controlled at this stage.**

If you are not able to control your mind, the few stray thoughts will begin to gain traction. You may start to admire the person in question. Your mind will be comfortable when they are with you. When you are apart, your mind will dwell on them and your sense organs will seek them out. It will gradually develop an attachment (*Sangam*) toward this person. Your intellect *must* step in at this point since this attachment will make it increasingly difficult to control your mind.

However, if your intellect is *still* unable to exert control, your mind will develop a continuous flow of thoughts focusing on that person. You will develop a strong desire (Kaamam) for them, and in time you will come to feel that you cannot live without them. Unless your intellect is strong, it will be extremely difficult for you to regain control over your mind at this time.

At this juncture, your desire will either be fulfilled or remain unfulfilled. If it *is* fulfilled, you will gain some temporary happiness. **However, since your mind will now be conditioned to the experience of developing and seeking out desires, it will simply move on to your *next* unfulfilled desire.**

If your desire is *not* fulfilled, then your continuous flow of thoughts will become obstructed. Obstructed thoughts lead to anger. When your mind becomes angry (*Krodham*), your intellect will recede, and you will enter a period of temporary madness or delusion (*Sam-Moham*) until you calm down.

The longer you spend in delusion, the greater the damage to your personality. As you languish in this condition, you will gradually lose your memory of all the knowledge you have acquired. Eventually, this will lead to the loss of your capacity to think—and, in time, to *Buddhi-Naasham*, or 'the total destruction of your intellect.'

Once your intellect is lost, your personality is totally destroyed (*Pra-Nashyam*). You will be reduced to the level of an animal. Without your intellect, you will no longer be qualified to pursue the four human goals (Purusha-Artham), which means that you will no longer be able to use this life to reduce your Sanchitham.

Thus, the loss of your intellect effectively disqualifies you from capturing infinity in this lifetime. This grave consequence highlights the importance of being prepared to use your intellect to keep your mind in check throughout your life.

How to stay on the path to the infinite

The fact that you have reached the end of this book demonstrates that your spiritual progress is already highly advanced. After an endless series of births and rebirths, you are finally on the cusp of capturing infinity. You must not allow yourself to lose this extremely rare opportunity to gain salvation (Moksham) and escape the never-ending story of reincarnation!

A strong intellect is your most important asset; it is the key to staying on the path of spiritual progress. To perform the prescribed yogas correctly, you must commit to strengthening your intellect.

If you wanted to excel in a sport or some other physical pastime, you would commit to developing and maintaining your muscles every single day. Similarly, you must commit to developing and maintaining your intellect every single day.

How do you develop a stronger intellect? By exercising your *existing* intellect regularly. Refuse to take anything for granted when you are presented with new information. Question everything. Use your intellect to verify and validate facts, lessons, and teachings for yourself. This will ensure that your intellect is always available to you, which is the key to purifying your mind and aligning it with the intellect—a prerequisite to capturing infinity.

Continuously seek out new sources of information to help grow your understanding and refine your intellect—particularly about Vedanta. While this book has presented the key concepts and methods involved in Vedanta, it is merely an introduction to the vast and deep Vedantic scriptures. It is recommended that you continue your study by seeking out additional resources.

Numerous Vedantic institutes across the globe are dedicated to teaching these scriptures. Information regarding some of these institutes will be included as a reference upon request by email. You are encouraged to seek out an institute that is suited to your geographical location, financial position, spoken language, learning style, and other life situations.

If you need guidance as to which institute may be most appropriate to your background and situation, the author can offer recommendations. The author may also be able to address any questions and clarifications you may have at this point, in addition to offering recommendations for guided mediation courses. Please email all communications to: **life@vedantic.org**

As you close this book, be mindful of the fact that the ideas presented here are simply representations of ideas enumerated by various masters in the Vedantic scriptures. Though the author have made deliberate changes to the sequence in which certain ideas are presented to appeal to the scientific community, the ideas themselves are akin to 're-tweets' of the timeless teachings. The author is also walking the same path that you are now on—the path towards infinity.

About the Author

Prabha is a renowned meta-physicist, engineer, and entrepreneur. He has dedicated over 25 years in study and research on Causality, Creation, Infinity, Consciousness, and Self-Development. He observed that many physicists, thinkers, researchers, readers, struggle to grasp Trans-Scientific concepts. Through his contemporary and accessible explanations, Prabha simplifies these trans-scientific mysteries, making them understandable for readers of all backgrounds.

By applying simple meta-physical methodologies in his engineering business, Prabha has achieved significant material success. Prabha is convinced that anyone can embark on this transformative journey and attain extraordinary accomplishments.He lives in Dubai with his wife, Rosh, who unconditionally supports all his endeavors.

www.ingramcontent.com/pod-product-compliance
Lightning Source LLC
Chambersburg PA
CBHW031336230926
51848CB00002B/260
9789696792642